Charles Barnard, Alfred Marshall Mayer

Light

A Series of Simple, Entertaining, and Inexpensive Experiments in The
Phenomena of Light, for The Use of Students of Every Age

Charles Barnard, Alfred Marshall Mayer

Light

A Series of Simple, Entertaining, and Inexpensive Experiments in The Phenomena of Light, for The Use of Students of Every Age

ISBN/EAN: 9783337266974

Printed in Europe, USA, Canada, Australia, Japan

Cover: Foto ©ninafisch / pixelio.de

More available books at **www.hansebooks.com**

NATURE SERIES.

LIGHT:

A SERIES OF

SIMPLE, ENTERTAINING, AND INEXPENSIVE EXPERIMENTS IN THE PHENOMENA OF LIGHT, FOR THE USE OF STUDENTS OF EVERY AGE.

BY

ALFRED M. MAYER

AND

CHARLES BARNARD.

WITH ILLUSTRATIONS.

London:
MACMILLAN AND CO.
AND NEW YORK.
1889.

RICHARD CLAY AND SONS, LIMITED,
LONDON AND BUNGAY.

First Edition, 1878; *Reprinted*, 1879, 1889.

PREFATORY NOTE.

IT is the design of this book to furnish a number of simple and easy experiments in the phenomena of light, that any one can perform with materials that may be found in any dwelling-house, or that may be bought for a small sum in any town or city. By the aid of this book the reader becomes an experimenter. The student of Nature may read in books, and soon forget. The experimenter who questions Nature himself, who constructs his own apparatus, and who performs his own experiments, learns past forgetting. He knows because he has observed.

It is believed that this book will occupy a place hitherto unfilled in scientific literature. It is specially prepared for the boy or girl student, and for the teacher who has no apparatus, and who wishes his pupils to become experimenters, strict reasoners, and exact observers. Nearly all the experiments described are new, and all have been thoroughly tested. The materials employed are of the cheapest and most common description, and all the experiments may be performed at an expense of less than £3. The apparatus is, at the same time, suitable for regular daily use in both the home and school, and with care should last for years.

The origin of this series of books, and the manner of their production, may be briefly stated. For several years Professor Mayer has been studying how to give to every teacher and scholar the knowledge of the art of experimenting. To accomplish this very desirable object, he had invented the simplest and cheapest apparatus, and he and his scientific friends had been satisfied with their performance. It remained to describe these instruments and the ways of using them. He found, however, that his leisure from professional duties was not sufficient for this work; and, not to delay further the publication of his labours, Professor Mayer called in Mr. Charles Barnard to assist him in preparing the books for the press. The construction and arrangement of the instruments were explained to Mr. Barnard, and the experiments were made before him. Mr. Barnard then wrote out the descriptions, which were revised by Professor Mayer. The engravings have been made under Professor Mayer's special direction, and care has been taken to render them accurate representations of the apparatus and experiments.

The nature of light is not touched upon in this volume. The authors propose to explain, in another book, the phenomena of interference and polarisation of light, and to explain fully the structure of the eye and the nature of vision.

THE AUTHORS.

CONTENTS.

	PAGE
PREFATORY NOTE.	v

CHAPTER I.

Introduction 1

CHAPTER II.

The Sources of Light 4
The Heliostat 7
First Experiment with the Heliostat 14
Experiment with Cards and a Lamp 16
Experiment with Shadows 19
Experiment in Measuring Light 23

CHAPTER III.

Reflection of Light 26
Experiment in Multiple Reflection 31
Second Experiment in Multiple Reflection 33
Experiment with Mirror on Pulse 34
Experiment with Glass Tube 36
Experiments in Dispersed Reflection 37
Experiment with Jar of Smoke 41
The Milk-and-Water Lamp 43

CHAPTER IV.

	PAGE
Refraction of Light	44
The Water-Lens	54
Experiments in Projection	59
The Fountain of Fire	62
The Water-Lantern	65
The Solar Microscope	70

CHAPTER V.

The Decomposition of Light	72
Experiments with the Solar Spectrum	75
The Colour-Top	78
Direct Recomposition of the Colours of the Spectrum	81
Experiments in Reflected Colours	81
Experiments in Contrasted Colours	83
CONCLUSION	88
LIST OF APPARATUS AND THEIR COST	90

LIGHT.

CHAPTER I.

INTRODUCTION.

ALL about us are men busy with their various trades and professions: sailing ships, digging in mines, making all manner of useful tools and machinery, planting seeds and reaping harvests, and doing many other works and labours according to certain fixed rules that they found printed in books, or that they learned of others, or that they discovered for themselves. Each one has to do with the physical phenomena around him. The more he knows about these phenomena—the more he knows about things, their relation to each other, and their action one upon another—the better he can work. A knowledge of the phenomena of Nature is the most important knowledge one can have who wishes to succeed in life. More than this, the observation of facts in Nature gives readiness of perception, and study and reading upon the causes of these facts stimulate the mind to healthful and pleasurable action.

The laws that govern the physical phenomena about us were not told to us by ancient gods or divinely-instructed men. They were discovered by experiment or observation. Men asked questions of Nature; they watched her phenomena till they felt sure they saw a reason for their action. Sometimes they did not understand all that happened, and made strange guesses at the laws that governed the happenings. Other men repeated the experiments and got new answers; and thus, in time, the truth about things became known. Many of these facts in Nature, and the laws that govern their action, are now known of all men. Others are still obscure, or dimly known, and are being investigated every day in the hope that they may be better understood.

The farmer, the sailor, the mechanic, and artisan, most familiar with these facts and laws of Nature, is, other things being equal, the most likely to be successful in his work. You hope to have a share in the world's work, and you wish to study Nature and her phenomena. You can read about these things in books. A better way is to make experiments—to ask Nature yourself—to examine the phenomena of light, heat, sound, electricity, etc., to study these phenomena, and find their causes for yourself.

To try an experiment means to put certain things in certain relations with other things, for the purpose of finding out how they affect each other. Experimenting is thus a finding out.

It is the design of this book to tell you something about experiments in the phenomena of light—to show how these experiments illustrate the action of light, and to explain briefly some of the elementary laws that govern this action. All of these experiments may be performed with the cheapest and most common materials that can be found. They are all easy and simple, and they are, at the same time, interesting and entertaining. Some of the things here described are capable of affording amusement for a large number of people, and many of the exhibitions and displays that may be made with them are wonderfully attractive and beautiful.

CHAPTER II.

THE SOURCES OF LIGHT.

WHEN the sun rises in the morning, the darkness of the night seems to fade away, and, wherever we look, without or within, all the air and space about us appears to be full of light. When evening comes again, the daylight disappears, and the moon and the stars give us another light. In the house we light the lamps, and they give us another light. Out-of-doors, in the dusky meadows, we see the fire-flies darting about, and giving out pale sparkles of yellow light as they fly. We look to the north in the night and see the aurora, or we watch the lightnings flash from cloud to cloud, and again we see more light.

This light from sun and moon, the stars, the fire, the clouds, and sky, is well worth studying. It will give us a number of the most beautiful and interesting experiments, and, by the aid of a lamp, or the light of the sun, we can learn much that is both strange and curious, and perhaps exhibit to our friends a number of charming pictures, groups of colours, magical reflections, spectres, and shadows. All light comes from bodies on the earth or in the air, or from bodies outside of the atmosphere; and these bodies we call the sources of light. Light from sources outside of the atmosphere we call celestial light, and the sources of this light are stars, comets, and nebulæ. The nebulæ appear like flakes and clouds of light in the sky, and the comets appear only at rare intervals, as wandering stars that shine for a little while in the sky and then disappear. The stars are scattered widely apart through the vast spaces of the universe, and they give out their light both day and night. The brightest of these stars is the sun. When it shines upon us, the other stars appear to be lost in the brighter light of this greater star, and we cannot see them. At night, when the sun is hid, these other stars appear. We look up into the sky and see thousands of them, fixed points of light, each a sun, but so far away that they seem mere spots and points of light. Besides these stars are others, called the planets, that move round the sun. These give no light of their own, and we can only see them by the reflected light of the great star in the centre of our solar system. Among these stars are the Moon, Venus, Mars, Jupiter, and many others. We might call celestial light starlight; but the light from the great star, the sun, is so much brighter than the light of the others, that we call the light it gives us sunlight, and the light from the other suns we call starlight. For convenience, we also call the reflected light from the planets starlight, and the light from our nearest planet we call moonlight.

Terrestrial light includes all the light given out by things on the earth, or in the air that surrounds the earth. The most common light we call firelight, or the light that comes from combustion. When we light a lamp or candle, we start a curious chemical action that gives out light and heat. The result of this action is fire, and the light that comes from the flame is firelight. When a thunderstorm rises, we see the lightning leap from the clouds, and give out flashes of intensely bright light. Sometimes, at night, the northern sky is full of red or yellow light, darting up in dancing streamers, or resting in pale clouds in the dark sky. You have seen the tiny sparkles of light that spring from the cat's back when you stroke her fur in the dark, or have seen the sparks that leap from an electrical machine. All these—the aurora, the lightning, and the electric sparks—are the same, and we call such light electric light.

Sometimes, in the night, we see shooting-stars flash across the sky. These are not stars, but masses of matter that, flying through space about the earth, strike our atmosphere and suddenly blaze with light. The friction with the air as they dart through it is so great that these masses glow with white heat, and give out brilliant light. Two smooth white-flint pebbles, or two lumps of white sugar, if rubbed quickly together, will give out light, and this light we call the light from mechanical action.

Sailors upon the ocean sometimes see, at night, pale-yellow gleams of light in the water. A fire-fly or glow-worm imprisoned under a glass will show, in the dark, bright spots of light on his body. A piece of salted fish or chip of decayed wood will sometimes give a pale, cold light in the night; and certain chemicals, like Bologna phosphorus and compounds of sulphur, lime, strontium, and barium, if placed in the sunlight in glass vessels and then taken into the dark, will give out dull-coloured lights. All these—the drops of fire in the sea, the glow-worm, the bit of decayed wood, and these chemicals—are sources of the light called phosphorescence.

These are the sources of light—the stars, the fire, electricity, friction, and phosphorescent substances. We can study the light from all of them, but the light from the sun or a lamp will be the most convenient. The light of the sun is the brightest and the cheapest light we can find, and is the best for our experiments. A good lamp is the next best thing, and in experimenting we will use either the sun or a lamp, as happens to be most easy and convenient.

THE HELIOSTAT.

In looking out-of-doors in the daytime we find that the sunlight fills all the air, and extends as far as we can see. It shines in at the window and fills the room. Even on a cloudy day, and in rooms where the sunshine cannot enter, the light fills everything, and is all about us on every side. Now, in studying light we do not wish a great quantity. We want only a slender beam, and we must bring it into a dark room, where we can see it and walk about it and examine it on every side, bend it, split it up into several beams, make it pass through glass or water, and do anything else that will illustrate the laws that govern it.

Choose a bright, sunny day, and go into a room having windows through which the sun shines. Close the shutters, curtains, and blinds, at all the windows save one. At this window draw the curtain down till it nearly closes the window, and then cover this open space with a strip of thick wrapping-paper. Cut a hole in this paper about the size of a halfpenny, and at once you will have a slender beam of sunlight entering the hole in the paper and falling on the floor. Close the upper part of the window with a thick shawl or blanket, and, when the room is perfectly dark, our slender beam of light will stand out clear, sharp, and bright.

As soon as we begin to study this beam of light, we find two little matters that may give us trouble. The sun does not stand still in the sky, and our beam of light keeps moving. Besides this, the beam is not level, and it is not in a convenient place. We want a horizontal beam of light, and some means of keeping it in one place all day. An instrument that will enable us to do this, and that can be adjusted to the position of the sun in the sky at all seasons of the year and every hour of the day, may be readily made, and will cost only a small sum of money.

On the next page are several drawings, giving different views of such an instrument and some of its separate parts. It is called a heliostat, and we shall find it of the utmost value in our experimenting in light, heat, sound, electricity, and other branches of physical science.

The first drawing represents a front-view of the heliostat. The second drawing gives an end-view, and we can now make one by simply following these few directions: The part marked A in the two drawings is a piece of pine board, 23 inches (58.4 centimetres) wide and two or more feet long, or as long as the window where it is to be used is wide. Any boy who can use plane and saw can make this piece of work out of common inch-board, and, if you have no pieces so wide as that, it can be made of two or more pieces fastened together with cleats; but, in this case,

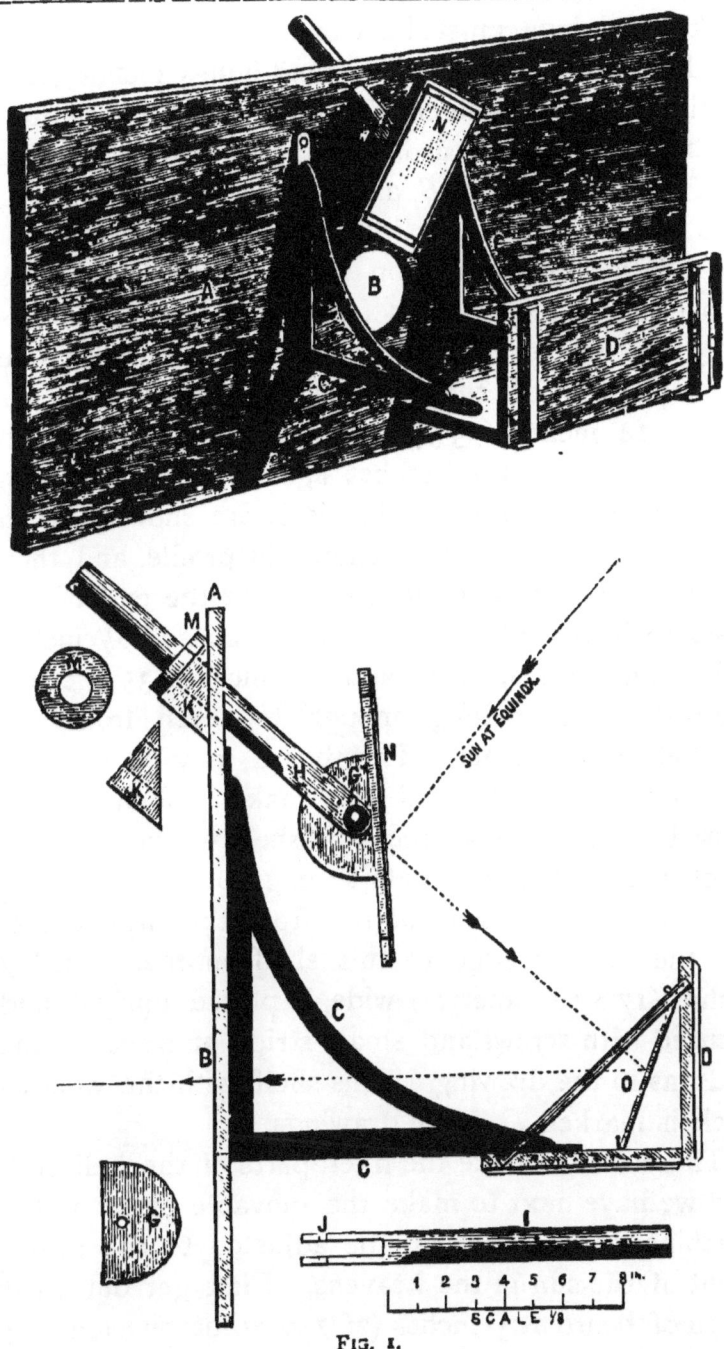

Fig. 1.

all the joinings must be close and tight. In the middle of this board, cut a round hole 5 inches (12·7 centimetres) in diameter, with its centre 8 inches from the bottom of the board. In the first drawing this hole can be seen at *B*, and in the second drawing it is shown by dotted lines at *B*. On one side of the board screw two iron brackets, using brackets measuring 14 inches (35·5 centimetres) by 12 inches (30·5 centimetres). These brackets are placed one on each side of the hole in the board, and are placed 14 inches (35·5 centimetres) apart, and with the short arm of the bracket against the board. In the first drawing the two brackets are shown, and in the second drawing one is shown in profile, and they are marked *C* in both drawings. On the end of the brackets is placed a flat piece of board, $6\frac{1}{2}$ inches (16·5 centimetres) wide and 14 inches (35·5 centimetres) long, or long enough to reach from one bracket to the other. This board may be screwed up to the brackets, and thus make a shelf. Care must be taken in fastening this shelf to the brackets to place it so that the outside edge of the shelf will be 16 inches (40·6 centimetres) from the large board. On the outside edge of this shelf another board, 7 inches (17·8 centimetres) wide, is placed upright, and secured with screws and small strips of wood at the ends, as in the drawing. This shelf, with the wooden back, is marked *D* in the drawings.

These things make the fixed parts of the heliostat, and we have next to make the movable parts, or the machinery whereby it can be adjusted to the movement of the sun in the heavens. First, get out a flat piece of board $10\frac{1}{2}$ inches (26·7 centimetres) long, $6\frac{1}{4}$

inches (16 centimetres) wide, and ½ inch (12 millimetres) thick. Then make a flat, half-round piece, shaped like the figure marked *G*. This piece must be ¼ inch (7 millimetres) thick, 5½ inches (14 centimetres) along the straight side, and with the circular part with a radius of 3 inches (7·6 centimetres). A hole ½ inch (12 millimetres) in diameter, is made in this, as represented in the drawing, and then the half-round piece must be screwed to the flat piece of wood we just cut out. In the figure marked *N* you will see these two pieces fastened together. Fig. *I* is the most difficult piece of all. It should be made of ash or some hard wood. One end is square, and has a deep slot cut in it; the rest is round, and may be 1½ inch (32 millimetres) in diameter. The square part must be large enough to slip over the half-circular piece, *G*, as is shown at *H*. A hole, ½ inch (12 millimetres) in diameter, is cut in the two ends, as marked by dotted lines at J, and through these holes an iron bolt and nut are fitted, so as to hold the circular piece, *G*, and yet allow it to turn freely in every direction. A hole, 1¼ inch (32 millimetres) in diameter, is cut through the triangular piece of wood *K*, as shown by the dotted lines, and then this block is securely fastened to the back of the large board, as shown in the second drawing. An opening of the same diameter, and having the same direction, is also cut through the board, and the movable piece, marked *I*, is put through this hole, as in the drawing. Finally, we want a wooden washer, 3½ inches (8·7 centimetres) wide, as represented at *M*. This we slip over the long wooden handle, as shown in the second drawing, and this washer rests on the block *K*, the top of which is 3½ inches square. This makes all the movable parts of the heliostat, and, when we have put in the mirrors, the instrument is finished and ready for use. We must have two mirrors, one 6 inches (15·2 centimetres) square and one 10 inches (25·4 centimetres) long and 6 inches (15·2 centimetres) wide. These may be made of common looking-glass; but plate-glass with silvered back is far better, and costs only a little more.

Any carpenter can make this instrument, and the cost will be about as follows: Wood, 2s.; labour, 7s.; glass, 4s.; iron nut and brackets, 2s.—total, 15s. When finished the instrument should have a coat of shellac varnish, and, when this is done, the mirrors may be put in place, and fastened on with very heavy bands of rubber. This will enable us to take the glasses off when the instrument is not in use, and, if the elastic bands or rings are very strong, they will answer perfectly. The long mirror is to go on the movable piece at N, and the small mirror stands on the shelf, facing the opening in the board, at O. This mirror stands at the angle shown in the next drawing (Fig. 2), and the other mirror is adjusted to the sun at its various positions in the sky at different seasons of the year.

Here is a diagram showing the position of the handle of the heliostat, and the mirror for different seasons and in different parts of the country. The handle must be placed on a line parallel with the axis of the earth, and the four dotted lines give its position when the heliostat is to be used in Boston, New York, Washington, and New Orleans. This also causes the block of wood marked K to have a slightly different shape, so that the hole through it will be in the middle. The dotted line marked "At Equinox" shows the path of the light from the sun, and the three dotted lines show the paths of the

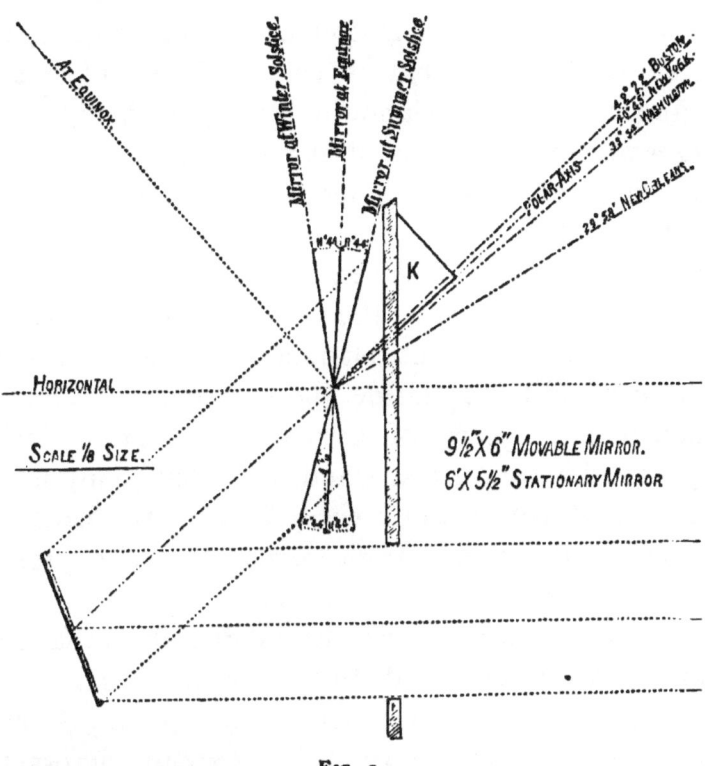

Fig. 2.

reflected light as it passes from one mirror to the other. The position of the movable mirror is also shown in the positions it has at summer and winter solstices.

FIRST EXPERIMENT WITH THE HELIOSTAT.

Choose a bright sunny day, and take the heliostat into a room having a window facing the south. Raise the sash and place the instrument in the window, and fasten it there so that it will be firm and steady. Before closing the window down upon it, move the larger mirror on its axis till it reflects a beam of light into the small mirror. Then turn the handle to the right or left, and a round, horizontal beam of light will enter the room. When this is done, close all the windows, so as to make the room as dark as possible. To do this, shawls or blankets or enameled cloth will be found useful inside the curtains and shutters. Then get a piece of cardboard about 6 inches (15·2 centimetres) square, and lay a halfpenny in the centre, and, with a knife, cut a hole in the card just the size of the coin. Then fasten this, with pins or tacks, over the opening in the heliostat.

We have now a slender beam of light in a dark room. Walk about and study it from different sides. See how straight this slender bar of light is; it bends to neither the left nor right, but extends across the room in an absolutely straight line. As the sun moves, turn the handle of the heliostat to keep the light in place.

Here is a picture of a dark room, in the window of which is the heliostat. In the centre of the piece of cardboard is the small hole where the light enters the room. A boy is holding one end of a long piece of linen thread just at the bottom of the hole in the card, and another boy has drawn the thread out straight and tight, so that it just touches the beam of light throughout its length.

Were you to try this experiment, you would see that the thread would suddenly be lighted up throughout its whole length, and would shine in the dark room like silver. Then if the boy allows the thread

FIG. 3.

to become slack and loose, or if he lowers it even a very little, it will disappear in the darkness. If he raises and lowers it quickly, it will seem to appear and disappear as if by magic.

This is a very pretty experiment; but we must not stop to look at its merely curious effects. Try it over several times, and see if it does not show you something about the beam of sunlight. Plainly, if the thread is lighted up its whole length when it is straight, then the beam of light must be straight also. Here we discover something about light; we learn that it has a certain property. Our experiment shows that light moves in straight lines.

EXPERIMENT WITH CARDS AND A LAMP.

Here is a picture representing three little wooden blocks placed in a row upon a flat, smooth table, and fastened to them are three postal-cards, so that they will stand upright. At the end of the table is a small lamp. This is all we need to perform another experiment, that will show us the same thing we observed with the beam of light from the heliostat. To make these things, get a piece of wood 10 inches (25·4 centimetres) long, 3 inches (76 millimetres) wide, and $1\frac{1}{2}$ inch (37 millimetres) thick, and saw it into five pieces, each $2\frac{1}{2}$ inches (64 millimetres) long. Next make three slips of pine, 4 inches (10 centimetres) long, 3 inches (76 millimetres) wide, and $\frac{1}{8}$ inch (4 millimetres) thick. Having made these, get three postal-cards, and lay them flat on a board, one over the other. Just here we need a tool for making small holes and doing other work in these experiments; and we push, with a pair of pliers, a cambric needle into the end of a wooden penholder, or other slender stick, putting the eye-end into the wood, and thus making a needle-pointed awl. Measure off one-half inch from one end of the top postal-card, and with the awl punch a hole through them all, just halfway from each side. Lift the cards up, and with a sharp penknife pare off the rough edges of the holes, and then run the needle through each, so as to make the holes clean and even.

Take one of these cards and one of the wooden slips, and put the card squarely on one of the wooden blocks and place the slip over it, and tack them both down to the block. This will give us the cards and blocks as shown in the picture. When each card is

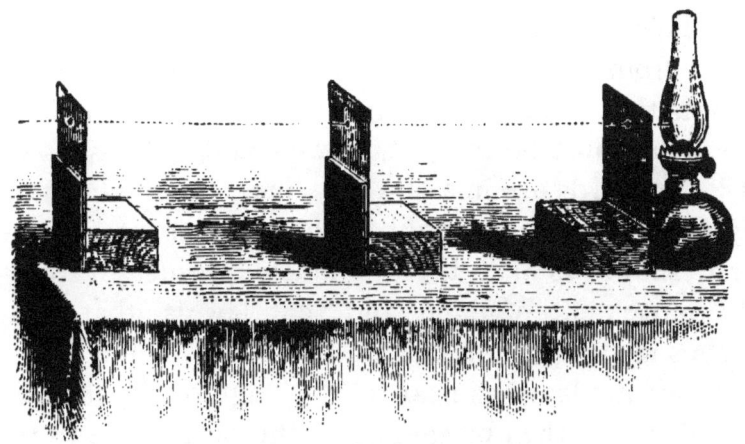

FIG. 4.

thus fastened to a block, we shall have two blocks left. These we can lay aside, as we shall need them in another experiment.

Now light the lamp, and place one block on the table, quite near the lamp. Look at the lamp carefully, and see that the flame is just on a level with the hole in the card. If it is too high or too low, place some books under it, or put the lamp on a pile of books on a chair near the table. Take a chair and sit at the opposite end of the table, and place another card before you. Now look, through the hole in this card, at the first card before the lamp. If the table is level, you will see a tiny star or point of light shining through the holes in the two cards. Without moving the eye, draw the third card into line between the others, and in a moment you will see the yellow star shining through all three cards.

Next take a piece of thread and stretch it against the sides of the three cards, just as they stand, and immediately you see that they are exactly in line. The holes in the cards we know are at the same distance from the edges of the cards, and our experiment proves that the beam of light that passed through all the holes must be straight, or we could not have seen it. The cards are in a straight line, and the beam of light must also be straight. This experiment, like the first, shows us that there is a law or rule governing the movement of light, and that law is, that light moves in straight lines.

Move the lamp as near to the edge of the table as possible, and then bring one of the cards close to the lamp-chimney. Then change your seat, and repeat this experiment several times in different directions. Each time you will see exactly the same thing, no matter in what direction the light moves from the lamp. The lamp may be moved from one side of the table to the other, and in every direction we shall find the light moving in exactly straight lines from the source of light. This is true whether the source be the sun, a lamp, or a star. One can walk all about the lamp and see it from every side, and we can place our three cards in any direction, north or south, up or down, east or west, or in any and every direction, and every time it will give the same result.

Thus we have found out the law by which light moves, viz., it moves in straight lines in all directions from the source of light.

Knowing this, you can readily think of a number of things in which these laws are made useful. A farmer planting an orchard, an astronomer fixing the positions of stars, a sailor steering his ship by night, employs this law: the first, to arrange his trees in straight lines; the second, to measure out vast angles in the sky; and the third, to lay the courses of his ship in safety. Each employs these laws with certainty and safety, because they are fixed and never change.

EXPERIMENT WITH SHADOWS.

This picture represents a sheet of white note-paper standing upright, like a small screen, upon a table. Near it is a bit of square paper, fastened to the end of our needle-pointed awl, and beside this is a lamp, and next to the lamp is a postal-card having a slit cut in it near the top. On the screen you will notice that there is a shadow of the bit of paper held on the needle. The paper screen may be made of any sheet of white paper, and it may be held upright by placing some books behind it. The bit of paper on the needle is just 1 inch (25 millimetres) square; and to hold the awl in place, the handle may be stuck in a mass of wax on the table. The slit in the postal-card should be 1 inch (25 millimetres) long and ¼ inch (7 millimetres) wide, and should be horizontal. The card may rest against the lamp, and, if it is not high enough, put something under it, so that the slit will be opposite the flame. These things are easily procured, and, when you have them, light the lamp, place the postal-card before it, and then make the room quite dark, or, if it is night, put out all the other lights. Set up the needle-awl with the bit of paper on the end about 12 inches (30·5 centimetres) from the lamp, and make it firm and steady with a bit of wax softened in the fingers. Then bring the screen in a line with the paper square and the lamp, and about

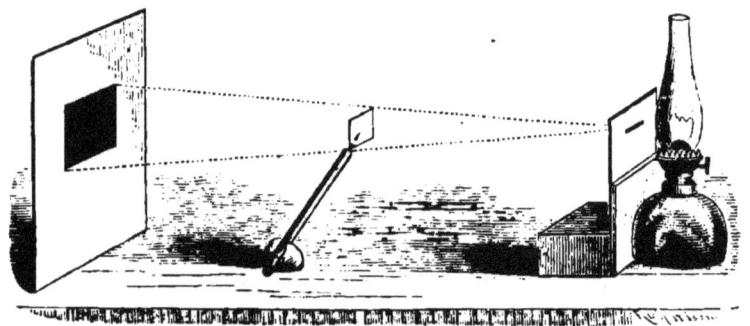

Fig. 5.

24 inches (61 centimetres) from the lamp. If everything is right, there will be a square shadow of the bit of paper on the screen. Look carefully at everything, and have the paper just on a level with the slit in the postal-card, and have the lamp, paper, and screen, just in line, and then the square shadow will appear sharp and clear on the white screen. With a lead-pencil trace an outline of this shadow on the screen; then move the screen just twelve inches (30·5 centimetres) farther from the lamp. Look at the shadow. See how it has increased in size. With the

pencil trace this shadow on the screen, and then, laying the screen on the table, measure the two shadows, and see how they compare in size, and see how they both compare with the size of the paper square that cast the shadows on the screen.

Fig. 6 shows how light spreads out, and how shadows expand as the distance increases. A is the lamp, and B is the postal-card, having a slit for the light to pass. C is the paper screen, and D is the first shadow made on the screen when it was 24 inches from the lamp. E is the second shadow made on the screen when it was 36 inches from the lamp. If you lay the paper C on the tracing of the small shadow D, you will observe that it only covers one-fourth of the surface, and that the shadow is four

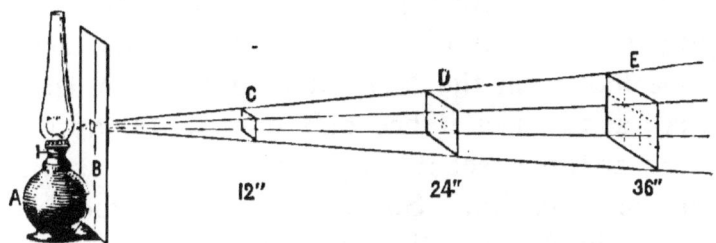

Fig. 6.

times as large. Place it on the larger shadow E, and you will see that it covers only one-ninth of its surface. In the diagram the first shadow is marked off into quarters, and the second into ninths, by dotted lines. The diagram also shows how the rays of light spread out wider and wider the farther they travel from the source of light.

Now, make two squares of paper, one the size of D and the other the size of E. Then place D 24 inches (61 centimetres) from the lamp, and E

36 inches (91·4 centimetres), and both in a line. If C, D, and E, have the positions shown in the diagram, it will be found that D and E are both in shadow, while the square C is illuminated. Remove the square C, and D will be lighted, which shows that all the light that falls on D previously fell on C. E yet remains in darkness. Next, remove D and replace C, and E still remains in shade; but, on removing C, E is fully illuminated. This shows that the quantity of light that fell on C spreads over four times the surface at the distance D, and nine times the surface at the distance E. Hence each one of the squares on D is one-fourth as bright as the square C, and any one of the squares on E is one-ninth as bright as C.

Here we are coming upon another fact about light; we find another law governing its action. At one foot from the lamp the light had a certain power; at two feet it had only one-quarter as much power; at three feet it only had one-ninth as much power or intensity. So, if we approach the lamp, at a certain distance the light has a certain brightness; at half that distance it has four times the brightness; at one-third the distance it has nine times as much brightness. The above relation existing between the intensity of the light on a surface at different distances from the source of light is often stated as follows: The illumination of a given surface varies in brightness inversely as the square of its distance from the source of light.

EXPERIMENT IN MEASURING LIGHT.

This picture represents a sheet of white paper standing upright upon a table. A few inches from this screen is our needle-pointed awl, stuck upright in the table. (If you do not care to do this, the awl can be stuck into a block of wood or bit of wax.) A lighted candle is placed on the table, about 22 inches (55·8 centimetres) from the screen. Beyond this is a lamp, placed upon a pile of books, so as to bring the flame of the lamp on a level with the flame of the

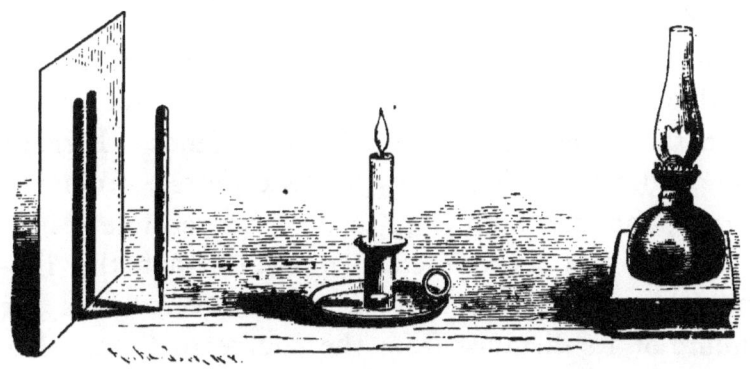

FIG. 7.

candle. The lamp should stand, say, at 44 inches (112 centimetres) from the screen; and, if it has a flat wick, it must be so placed that the wick stands diagonally to the screen.

These are all the things we need to make a most interesting experiment in measuring light, and we have only to make the room dark, or put out all the other lights if it is evening, and we can go on with the work. Let the candle burn a moment or two, and then bend the wick down, so as to give a large flame. If you have no lamp, a gaslight will answer. Upon the paper screen are two shadows of the awl side by side. Move the lamp to the right or left till the two shadows just touch, and make one broad band. Study this double shadow carefully. Perhaps one half is darker than the other. Move the lamp backward or forward, and you will see that its shadow changes—becomes darker or lighter. Presently you will find a place for the lamp where the double shadow appears of a uniform depth.

Now both lamp and candle cast just as deep a shadow, and yet one is much farther from the screen than the other. Measure off the distance. Perhaps the candle is 22 inches (55·8 centimetres) from the screen, and the lamp is 44 inches (112 centimetres).

In our last experiment we found that the illumination of a given surface varies inversely as the square of its distance from the source of light. The square of 22 is 484, and the square of 44 is 1,938. Now, if we divide 1,938 by 484, we get 4, and thus we find that our lamp is four times as bright as the candle. It casts just the same depth of shadow on the screen as the candle, and it is four times as bright, because the square of the distance of the candle will divide the square of the distance of the lamp four times. If we measure it another way we find the candle is half the distance from the lamp to the screen, and gives only one-quarter as much light.

Such a measurement as this is both easy and simple, and by means of such an experiment we can find out how much light any lamp gives. In this case, we find one lamp gives just four times as much light as the candle, or as much light as four candles would give at once. This is called a photometric experiment, from two Greek words meaning *light-measurement*. You may sometimes hear people say that a certain gas-lamp gives a sixteen or eighteen candle light, and our experiment shows us what they mean by this expression. They mean that the lamp has a photometric value of so many candles, or gives a light equal to the light of sixteen or eighteen candles burning at the same time.

CHAPTER III.

REFLECTION OF LIGHT.

PLACE the heliostat in position, and bring a slender beam of light into the darkened room. Then get a small looking-glass, or hand-mirror, and a carpenter's steel square, or a sheet of stiff paper, having perfectly square corners. Hold the mirror in the beam of light. At once you see there are two beams of sunlight, one from the heliostat and another from the mirror. Hold the glass toward the heliostat, and you will see this second beam going back toward the window.

This is certainly a curious matter. Our beam of light enters the room, strikes the mirror, and then we appear to have another. It is the same beam thrown back from the glass. This turning back of a beam of light we call the reflection of light.

Place a table opposite the heliostat, and place the mirror upon it, against some books. Turn the mirror to the right, and the second or reflected beam of light moves round to the right. Turn the glass still more, and the beam of light will turn off at a right angle, and there will be a spot of light on the wall at that side of the room. Now bring the carpenter's square or the piece of square paper close to the mirror, so that the point or corner will touch the glass just where the sunlight falls upon it. Now one edge of the square is brightly lighted by the sunbeam, and if the mirror is placed at an angle of 45 degrees with the sunbeam, the other edge of the square is lighted up by the second beam.

In this diagram, A is the beam of light from the heliostat, and B is the beam reflected from the mirror, that is marked M. To make this more simple, we call the first beam the beam of incidence, and we say

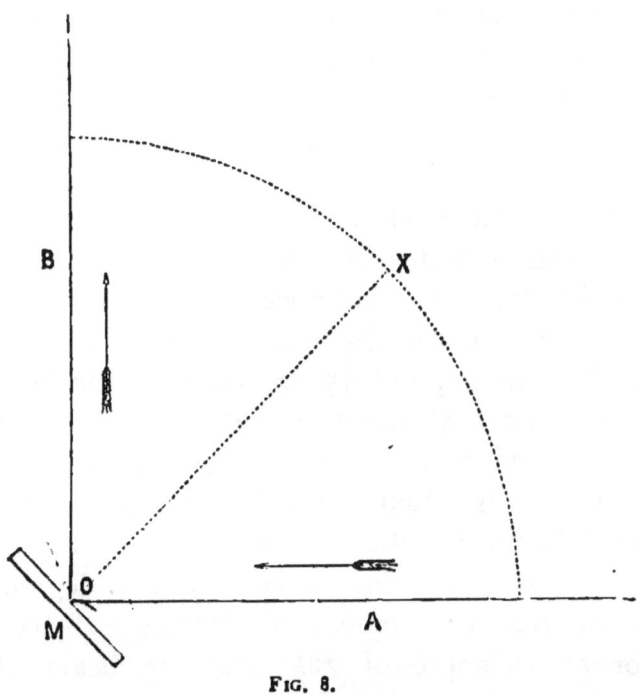

FIG. 8.

that it travels in the direction of incidence, as shown by the arrow. The second beam, marked B, we call the beam of reflection, and the course it takes we call the direction of reflection. The point marked O, where the light strikes the mirror, is called the point of incidence.

In the diagram is a dotted line representing a

quarter of a circle reaching from the beam of incidence to the beam of reflection. A quarter of a circle, as you know, is divided into 90 degrees. Another dotted line extends from O at the mirror to X on the quarter-circle, and divides it into two parts. Half of 90 is 45, and hence the mirror stands at an angle of 45 degrees with both beams of light. Now the line A and the dotted line reaching from O to X make the angle of incidence, and the angle between B and the line from O to X is the angle of reflection; and the curious part of this matter is, that these two angles are always equal. Here they are both angles of 45 degrees.

Move the mirror about in any direction, and measure the angles of incidence and the angles of reflection, and these angles will always be exactly equal.

If you look at the diagram you will see that the mirror is at an angle of 45 degrees with the beam of incidence, and that the beam of reflection is at an angle of 90 degrees with the incident beam. Hence, if the mirror is tilted through a certain angle, the reflected beam is tilted through twice this angle. For instance, if the mirror is moved 1 degree the beam of reflection moves 2 degrees. Place the mirror at an angle of $22\frac{1}{2}$ with the beam of incidence, and the beam of reflection is at an angle of 45. Move the mirror to an angle of $67\frac{1}{2}$, and the beam of reflection will move round to an angle of 135 degrees.

Fig. 9 represents the two postal-cards fitted on blocks of wood that we used in a former experiment, and the three blocks of wood we cut out at that time. The five blocks are placed close together in a line, and with the postal-cards at the ends. A lighted lamp is placed near one of the cards, and on the middle block is a small piece of window-glass that has been painted with black varnish. A single coat of black varnish on one side of the glass is all that is required to give us the black mirror needed in this experiment. Place the lamp close to the card in such a position that the flame will be just on a level with the hole in the card. If the lamp is not

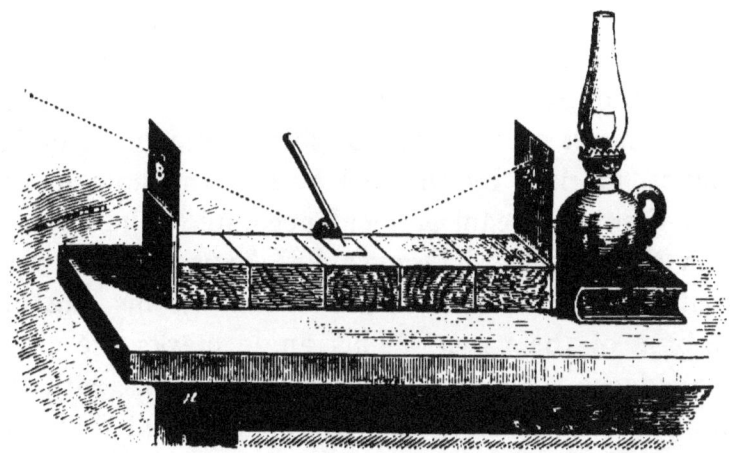

FIG. 9.

convenient, the blocks and cards may be placed upon a table facing a north window in full daylight.

When everything is ready, look through the hole in the postal-card marked *B*, down upon the black mirror, and on it you will see a single spot of light, the reflection from the lamplight or the light from the window shining through the hole marked *A* in the drawing. Get the needle-pointed awl and place it so that the point will just touch the spot of light in the black mirror, and then fasten the awl in this position with a piece of wax, as represented in the picture.

You will readily see that this experiment is the same as the last. Again we have a beam of light reflected from a mirror. The beam of incidence passes through the postal-card at A and finds its point of incidence on the mirror, and the beam of reflection extends from the point of incidence to the second card at B.

Take a sheet of stiff paper, 10 inches (25·4 centimetres) long, and about 4 inches (10 centimetres) wide, and hold it upright between the two cards, with the bottom resting on the mirror. With a pencil make a mark on the edge of this at the point of incidence, marked by the awl, and at the hole in the card where the beam of incidence enters, and marked A in the drawing. Draw a line between these two points and you have an angle formed by this line and the base of the paper. This angle marks the angle of incidence. Put the paper on the blocks, with the ruled line towards the card B, and you will find that the line fits here equally well. It now extends from the point of incidence to B, and proves that this angle is the same as the other, that both sides are alike, and that the angle of incidence and the angle of reflection are equal.

Take out the block in the middle, and move the others nearer together till they touch. Repeat the experiment: make a measurement with a piece of paper as before, and draw a line on it from the point of incidence to either of the holes on the cards, and then compare the angles thus found, and in each case they will be exactly alike. Take out another block and try it again, and you will reach the same result.

These experiments show us that there is a fixed law in this matter, and the more we study it the more we are convinced that it has no exceptions.

EXPERIMENT IN MULTIPLE REFLECTION.

Choose a south room on a sunny day, and close the blinds and shutters at all the windows save one, and at this window draw down the curtain until only a narrow space is left at the bottom. Close this space with a strip of thick wrapping-paper, and then cover the rest of the window with a blanket or shawl, so as to make the room perfectly dark. Then cut a round hole, the size of a halfpenny, in this paper, and through this hole a slender beam of sunlight will fall into the darkened room.

Bring a hand-mirror into this beam of light, and the beam of reflection will make a round spot of sunlight on the wall above the window. This spot of light is a picture of the sun thrown by the mirror upon the wall. Hold the mirror at an oblique angle in the sunbeam, and direct the beam of reflection upon the opposite wall. Now there are several reflections, brilliant spots of light. If the spots of light do not stand out sharp and clear, turn the mirror slowly round and you will soon find a position for the glass that will give six or more reflections.

How does it happen that a common looking-glass can thus split a single sunbeam into several beams? If you put a pencil to a mirror you will notice that while the point of the pencil touches the glass the point of the reflected pencil seen in the mirror does not meet the point of the real pencil, and that there is a little space between them. The reflection we see in the glass is from the smooth surface of the quicksilver at the back of the glass, and the space between the reflection and the pencil is filled by the glass.

Hold a sheet of common window-glass before a lighted lamp or candle, and you will see a faint reflection of the flame in the glass, and at the same time you can readily see through the glass. This shows us that the outside of any piece of smooth glass will reflect light, and our experiment is designed to show a still more curious matter.

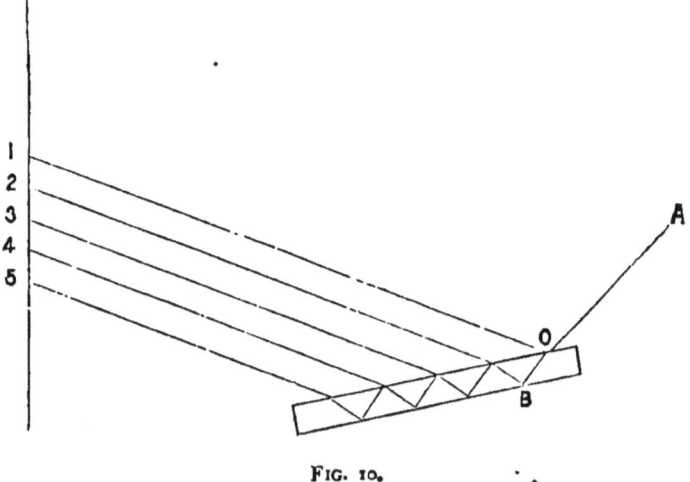

FIG. 10.

Fig. 10 represents the single beam reaching the point of incidence on the outside of the mirror at O, and reflected to the wall at 1. Part of the light goes through the glass to B, and here is another point of incidence, and a new beam of reflection is thrown through the glass to the wall at 2. If you look at the reflections on the wall, you will see that the second spot of light is the brightest. This comes from the quicksilver, for, as this is a better reflector than the glass, it sends out a brighter beam of reflection. When this second beam of reflection passes through the glass, a part of its light is reflected from the under side of the surface, and is turned back against the quicksilver again. Once more it is reflected, and a new beam of reflection makes number 3. The drawing shows the path these beams of light take in the glass, and the quivering spots of light on the wall show how one beam of light may be reflected again and again in different directions. If the reflector was perfect and returned all the light, these multiple reflections might be repeated many times over; but every time light is reflected from any bright surface, a part of the light is lost, and thus each reflection grows fainter and fainter till the light is spent. Look at the multiplied reflections on the wall, and you will see that the first reflection from the glass is bright, and that the second, from the quicksilver at the back of the glass, is brighter still; and that the others grow fainter and fainter till all the light is spent, and the reflections disappear.

SECOND EXPERIMENT IN MULTIPLE REFLECTION.

Light a lamp and place it on a table, and get the two postal-cards and the blocks that we used in the experiment in reflection. With a sharp knife cut a slit in one card, just at the pin-hole, about $\frac{3}{4}$ inch (19 millimetres) long and $\frac{1}{25}$ inch (1 millimetre) wide. Then place this card close to the lamp, as in the other experiment, and set up the other card about fifteen inches away from it. Then lay a looking-glass on the table between the two. Look at the picture (on page 29) and arrange the cards as there represented, and put the mirror in place of the blackened glass on the blocks. On looking through the small hole in the postal-card (marked B in the drawing), you will see in the mirror several bars of yellow light, placed one over the other. Again we have an instance of multiplied reflection. Instead of seeing the reflections thrown upon the wall, we can look down upon them and see them, just as they stand, each at its point of incidence on the glass and the quicksilver. Study these brilliant bars of light, examine the diagram carefully, and you will readily see that this experiment simply exhibits in a different manner the same thing that we saw in the last experiment.

EXPERIMENT WITH MIRROR ON PULSE.

Get a small bit of looking-glass, about an inch (25 millimetres) square, and some wax. Warm the wax in the hand till it is soft, and then make three small pellets about the size of a pea. Put one of these on the back of the little mirror, near the edge and halfway between two corners. Place one at each of the opposite corners, so that the mirror will have three legs or supports placed in a triangle. Put the heliostat in place, and bring a small beam of sunlight into the dark room. If this is not convenient, any beam of sunlight in a dark room (as in former experiments) will answer.

Turn back your coat-sleeve, and, while standing near the beam of light, place the little mirror on the wrist, with one of the wax legs resting on the pulse. Then bring the arm into the beam, so that the light will fall on the mirror. Hold the arm steady, and watch the spot of reflected light thrown upon the wall. See! It moves backward and forward with a curious, jerking motion. It is like the ticking of a clock, or like the beating of one's pulse. It is the motion of your pulse. The mirror moves with the pulse, and the beam of reflection thrown on the wall moves with it, and, though this movement is very slight, the reflection on the wall moves over a space of several inches, and we can see it plainly. In our first experiment in reflection we learned that when a mirror was moved to the right or left, the beam of light reflected from it moved also to the right or left, and each time through twice as great an angle as the mirror.

This experiment is a wonderfully interesting one, and may be tried with a number of boys or girls, and each may see the peculiar beating of his or her pulse pictured on the wall in the most singular and startling manner. If any of the persons whose pulse-beats are thus exhibited get excited, laugh at the exhibition, or are in any way disturbed, the change in the movement of their pulse will be quickly repeated on the wall, where a hundred people can see it.

EXPERIMENT WITH GLASS TUBE.

Procure a glass tube, about ¾ inch (19 millimetres) in diameter and 12 inches (30·5 centimetres) long, and paint the outside with black varnish. If this is not convenient, cover the tube with thick black cloth, and fasten it down with mucilage, taking care to have the cloth square at the ends. Punch a hole in a postal-card with the sharp point of a pair of scissors, and with a knife make the ragged edges of the hole

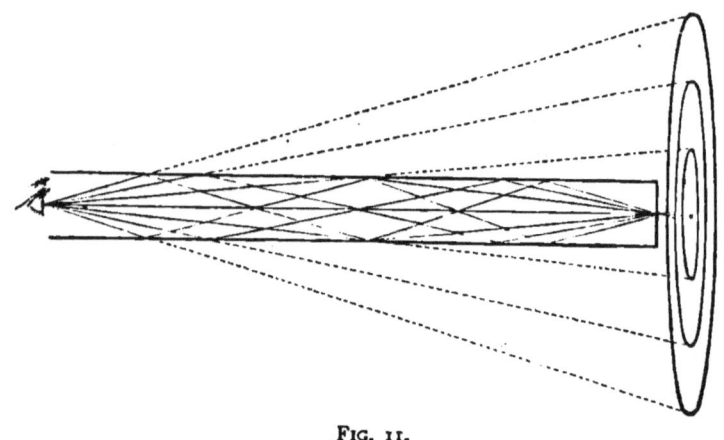

FIG. 11.

smooth. Hold the card at one end of the tube so that the hole will come just at the centre of the opening, and then, while facing a window or a bright lamp, look through the tube with one eye, and you will see a spot of light surrounded by a number of beautiful rings.

Here we have another example of multiplied reflection. The light entering the tube, through the hole in the card, falls on the smooth surface of the interior of the tube, and appears to the eye in the form of rings.

Fig. 11 represents a section of the tube, and shows the paths the different rays of light take, and shows how each is reflected from side to side till they all meet in the eye. The dotted lines and the rings projected beyond the tube show how they appear to the eye. By studying this drawing carefully, and trying cross cuts and slits in the card in place of the single hole, you will get a very correct idea of repeated reflection, and find the tube a source of considerable amusement.

EXPERIMENTS IN DISPERSED REFLECTION.

Get a small piece of black velvet or cloth and take it to a dark room, where the heliostat will give us a slender beam of sunlight. If this is not convenient, use a common beam of sunlight in a dark room, as in some of our former experiments. Hold the velvet in the hand between the fingers, and so as to leave the palm of the hand clear. Turn back the coat-sleeve so as to expose part of the white cuff, and then bring the velvet into the beam of sunlight. You will observe nothing in particular, for the black, velvet does not reflect the light at all. Now move the hand, so that the spot of light will fall on the palm. See what a pretty, rosy glow of light falls on the wall! This is the reflected light from the hand. The skin is rough, and the light is diffused and scattered about, and, instead of a bright spot of reflected light, as with a mirror, we have this glow spread all about on the wall and furniture. Now move your hand, so that the sunlight falls on your cuff. Immediately there is a bright light shining on the wall, and lighting the room with a pale, bluish-white glare. Move the hand quickly, so that the black cloth, the hand, and the white cuff, will pass in succession through the beam of light. Observe how the different things reflect the light in different degrees. The cuff is the smoothest and whitest, and gives the brightest reflection; the hand gives less light, because it is less smooth; and the cloth, that has a very dark and rough surface, gives no reflection at all, and the spot of sunlight falling upon it seems dull and faint.

This experiment shows us something more in the reflection of light. A piece of glass, the surface of water, polished metals, ice, and all substances having very smooth surfaces, reflect light in one direction. The linen cuff also reflected light, but apparently in a very different manner from the mirrors we have been using.

Place a lighted lamp upon a table and lay a mirror before it, and you can see a clear and distinct reflection of the lamp and the flame pictured on the glass. Put a sheet of white paper before the lamp, and you can see only a confused spot of reflected light on the brightly-lighted paper. Lay a freshly-ironed napkin or handkerchief before the lamp, and even the indistinct spot of light has disappeared, and the white cloth reflects light equally from every part.

These drawings are intended to show how light is reflected from different surfaces. The first represents a smooth surface, like glass, that sends all the beams in one direction, because the points of reflection for the beam are in the same plane. (See 1, 2, 3, Fig. 12.)

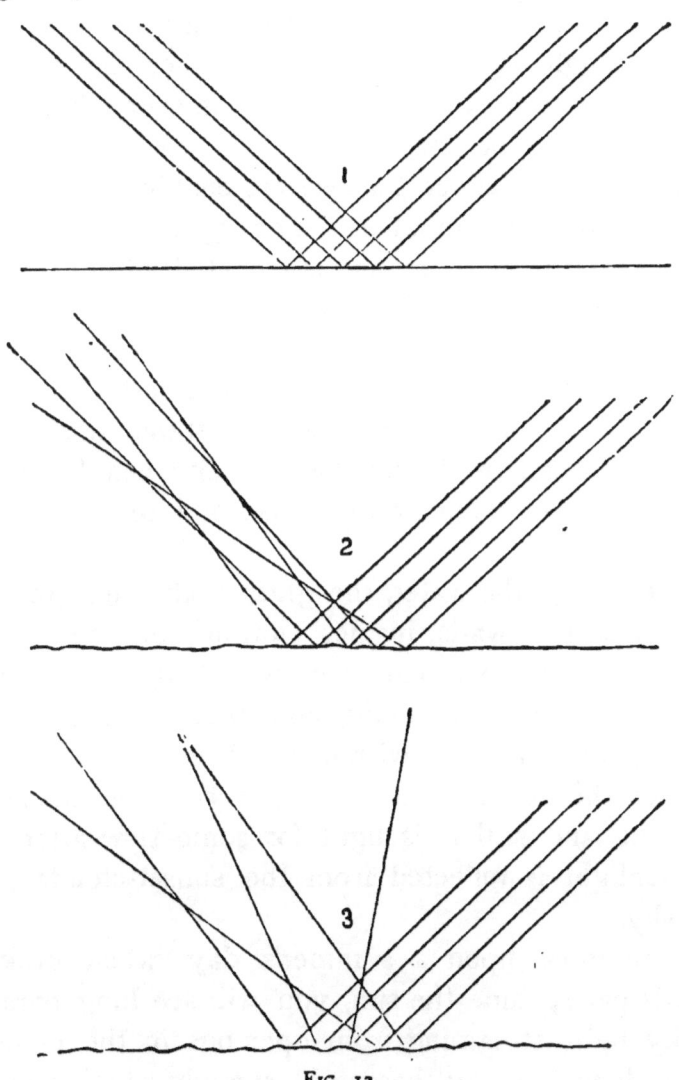

FIG. 12.

The second drawing represents a slightly-roughened surface, like paper. Some of the points of reflection turn the light one way, some another, and the beam of reflection is no longer formed of parallel rays. They are scattered about, and the image they form is confused and indistinct. In the third drawing we have a rough surface, like cloth, and here the rays of the beam of reflection are scattered in every direction, and we can see no image.

It is in this manner that we are enabled to see the people and things about us. The light of the sun or a lamp falls upon them, and is reflected into our eyes, and we say we see the objects. Very few things reflect light so brightly that we obtain from them a reflected image of the source of the light, and we generally see only dispersed and scattered light, that does not blind or dazzle the eye, and enables us to look upon these objects with ease, and to readily see all their parts.

The clouds, the water, the grass, rocks, the ground, buildings, the walls inside, clothing and furniture, and everything we can see, reflect light in every direction again and again, and thus it is that all spaces, without, and within, are filled with light so long as the sun shines. At night the sun sinks out of sight, and still it is light for some time after, for the sunlight is reflected from the sunset-clouds and the sky.

Sometimes, upon a summer's day, when broken clouds partly hide the sun, you will see long bars of dusky light streaming from openings in the clouds. These long bars are beams of sunlight shining upon dust and fine mist floating in the air, and we see them because each speck and particle reflects light in every direction.

EXPERIMENT WITH JAR OF SMOKE.

Fig. 13 represents a large, clean glass jar, such as one sees at the confectioner's. It is standing upon a black cloth laid upon a table in a dark room, and on top of the mouth is laid a postal-card, having

Fig. 13.

a slit, 1 inch (25 millimetres) long and $\frac{1}{25}$ inch (1 millimetre) wide, cut in it. Above the jar is a hand-mirror, so placed that the beam of sunlight from the heliostat (or from a hole in the curtain) will be reflected downward upon the postal-card on top of the jar.

This simple apparatus is designed to show how light is reflected from small particles floating in the air. Set fire to a small bit of paper and drop it into the jar. Place your hand over the mouth of the jar, and in a moment it will be filled with smoke. When the paper has burned out, put the postal-card in place, so that the slit will be in the centre of the mouth of the jar. Let the beam of reflected light from the mirror fall on this slit.

Look in the jar and you will see a slender ribbon of light extending downward through the jar. Elsewhere it is quite dark and black. Here we see the light streaming through the opening in the card, and lighting up the particles of smoke in its path.

Take off the card, and let the reflected beam fall freely into the jar. The smoke is now wholly illuminated, and the jar appears to be full of light, and every part of the bottle shines with a pale-white glow.

Put the postal-card on again and let the light fall through the slit. The smoke has nearly all disappeared, and the ribbon of light in the jar is quite dim. Curious streaks and patches of inky blackness run through it. What is this? Nothing—simply nothing. The smoke is melting away, and the beam of light disappears because there is nothing to reflect it and make it visible.

This part of the experiment appears quite magical in its effects, and is exceedingly interesting.

THE MILK-AND-WATER LAMP.

Take away the jar and put a clear glass tumbler in its place. Fill this with water and throw the beam of reflected light down upon it, and the water will be lighted up so that we can easily see the tumbler in the dark. Now add a teaspoonful of milk to the water and stir them together. Throw the beam of light down once more. This is indeed remarkable. The tumbler of milk-and-water shines like a lamp, and lights up the room so that we can easily see to read by its strange white light. Move the mirror and turn aside the beam of light, and instantly the room becomes dark. Turn the light back again, and once more the glass is full of light.

Here the minute particles of milk floating in the water catch and reflect the light in every direction, so that the entire goblet seems filled with it, and the room is lighted up by the strange reflections that shine through the glass.

CHAPTER IV.

REFRACTION OF LIGHT.

CERTAIN things, like glass, water, mica, and ice, allow light to pass directly through their substance. We hold them before the eyes, and see the light very nearly as well as through the air. Such substances, we say, are transparent. Other objects, like porcelain or oiled paper, do not permit all the light to pass, and such things we say are semi-transparent or translucent. Many other things do not permit light to pass through them, and cast shadows behind them when brought into a beam of light. These things cut off all the light, and we call them opaque.

Here is a common glass bottle with straight sides and about three inches (76 millimetres) broad, or as wide as a postal-card (Fig. 14). On one side is pasted a piece of white paper having a perfectly round hole cut in it. On the glass, in the clear space made by the circular opening in the paper, are two lines drawn at right angles, in ink. These two lines divide the circle into four equal parts, and are to serve as guides in some new experiments.

Fill the bottle with clear water up to the horizontal line in the circle, and then, holding the bottle in a small horizontal beam of sunlight from the heliostat, you will see that the light passes directly through the water in the bottle or through the air above the water. To make this more distinct, cut a slit, $1\frac{1}{2}$

inch (38 millimetres) long and $\frac{1}{25}$ inch (1 millimetre) wide, in a postal-card, and place this against the side of the bottle, so that the light will pass through

FIG. 14.

the slit. This gives a sharp, clear beam of light, and by studying it carefully, we see that the beam in the air and its continuation in the water preserve the same direction. If we place the bottle on the floor or table, and with the mirror send a perpendicular beam down into the water, we shall see exactly the same thing.

Fig. 15 represents the bottle of water standing upon a table, under a window, where the beam of sunlight enters from the heliostat. The opening

FIG. 15.

where the light comes in, the mirror, and the reflected beam of light thrown down upon the bottle, are plainly shown in the picture. The postal-card is held in such a position that the beam falls upon the slit and then enters the bottle. Look into the bottle through the opening in the paper, and see where the beam falls, and then move the mirror and the card till the beam enters the bottle above the water and strikes the water just where the two lines meet in the centre of the circle. Draw the postal-card forward so that some of the light will cross the outside of the bottle, and appears to make a white mark across the paper circle. Study the two beams outside and inside the bottle, and see if you can discern anything peculiar about them. The part of the beam inside the bottle and above the water follows the same direction as the beam outside till it touches the water-line, and then it turns down and takes a new direction. This bending, that takes place when a beam of light passes from air into water, is called refraction. It takes place very generally when light passes from one transparent medium to another, and gives rise to a number of curious matters in regard to light.

On p. 48 is a drawing of the beam of light crossing the opening in the paper, and showing how it is bent. It passes through the air above the water, in the upper half of the circle, and then takes a new direction through the water in the lower half. You will observe in the drawing dotted horizontal lines extend'ng from B to A, and from C to D. Look at the beam of light carefully, and with a pen mark these places A and C on the edge of the paper circle. Take the bottle to the light and measure off the distances from A to the perpendicular line, or along the line $A B$ in the drawing, and from C to the perpendicular line, or the line $C D$ in the drawing. Make a record of these measurements, and then take the bottle to the dark room. Place it nearer to the mirror, and let the reflected beam of light fall upon it at a different angle, being careful that the beam strikes the water at the centre of the circle. Examine the beam of light in the bottle, and you will observe that it is bent, but at a different angle. Mark the two points where the beam crosses the circle above and below the water, and measure their distances from the perpendicular line, and then compare these distances with those we obtained the first time. Divide the distance between A and B by the distance

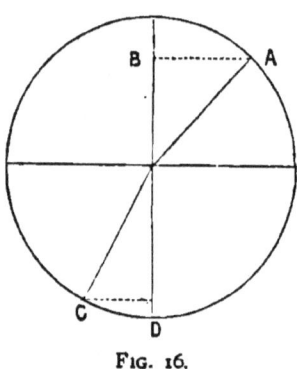

FIG. 16.

between C and D, and you will obtain a certain quotient. Divide the two sets of figures obtained the second time (that is, the distance from the edge of the circle to the perpendicular line above by the same below the water), and the quotient or ratio of the one to the other will be exactly the same as before. For instance, if the distance from A to B is four units, the distance from C to D will be three units, and in every experiment this proportion will be the same. In this case, where the light passes from the air to the water, we get a quotient that is one and a third, and this quotient we call the index of refraction. These experiments show us that there is a fixed law of refraction. When the light met the surface of the water at a right angle, it passed through the water without bending. In such instances the light is said to meet the water in a normal direction. If it meets the water on either side of this normal, it is refracted. Glass, diamonds, mica, and

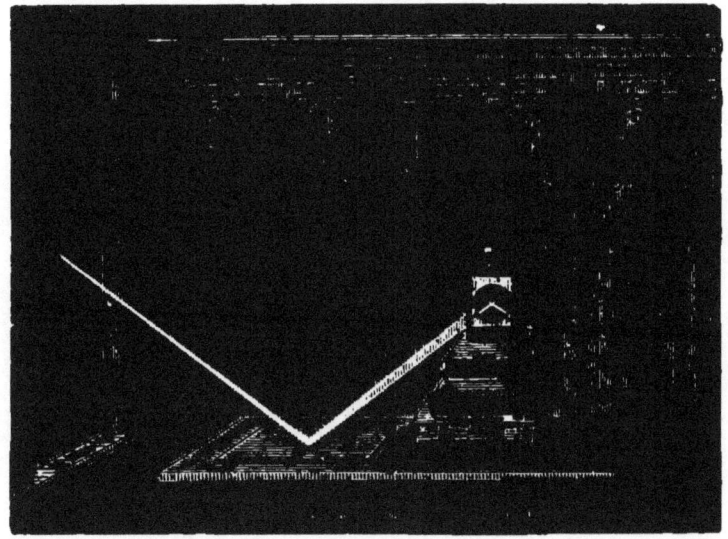

Fig. 17.

every transparent substance, have their own peculiar refraction. Glass has an index of refraction of 1·5. A diamond has quite another index of refraction, and it is by comparing these that we are able to prove whether a stone is a real diamond or only an imitation made of glass.

Above is a picture representing the bottle in a new position. The beam of sunlight enters the darkened window, and falls upon a mirror lying flat on the table. It is then reflected upward toward the bottle that stands upon a pile of books. The postal-card is put up as before, and the beam of light passes through the slit and enters the bottle below the surface of the water. Look at the beam of light in the bottle through the circular opening. Instead of passing through the water into the air above the surface, it is bent and turns downward into the water

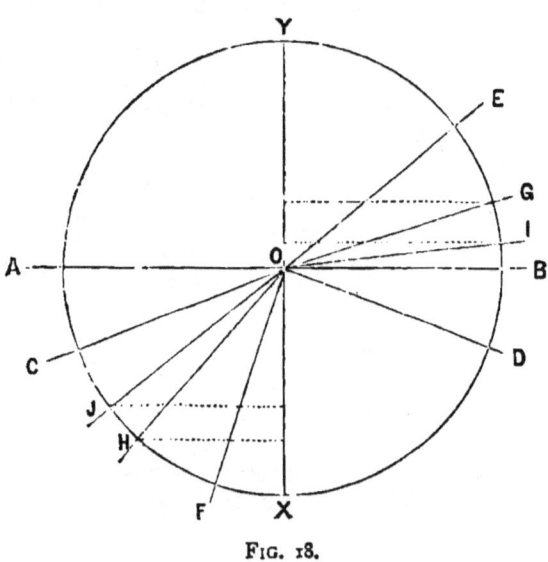

Fig. 18.

again. If, at first, you do not see this curious effect, raise the mirror slightly, tip it up toward the bottle, and take out some of the books under the bottle till the beam of light enters the bottle, in the direction $C O$, as in the above drawing. Here the line $A B$ represents the surface of the water in the bottle, and the line $Y X$ is the perpendicular line in the circle. In this experiment the light must enter the bottle at

C and pass to O at the surface of the water, and then you will see a most curious phenomenon, the reflection of the light from the surface of the water at O downward to D. To understand this singular matter we must study the diagram.

In the diagram a beam of light is represented as entering the circle at G, and is then refracted to H. Another beam goes from I to J. Dotted lines are drawn from each of these beams to the perpendicular, both above and below the water. You can easily compare the relations of these dotted lines above and below, and you will see that they still preserve the same relation to each other that we discovered in former experiments. First, we must observe that light may pass from air to water, as from G to O and H, or from water to air, as from H to O and G, and the amount of bending will be the same in both cases. In other words, the light takes the same path in going from air to water as when moving from water into the air. A beam of light passing to O, just above the surface of the water, will be refracted as already described. To study this matter further, we must reverse the direction of the light and cause it to pass from the water to the air. The beam of light entering at C, below the water, passes to O. Now, if we measure a line from C to the perpendicular OX, we shall find it is too long to be three units, if we call the length of the longest line (OB) that can be drawn from the circumference of the circle to its centre four units. The beam of light entering the water passes to the surface at O, and finds itself a prisoner, and it turns back and dives down into the water again. None of our experiments

E 2

have shown a more singular result than this. The lines which we have so often drawn perpendicular to the diameter, YX, of the circle, are called sines, and the law of refraction is always thus stated: The sines of the angles of incidence have a constant relation to the sines of the angles of refraction. In the case of light passing from air into water, the ratio of the sines is as 4 to 3; in the case of light passing from air into glass, the ratio of the sines is as 3 to 2.

The beam of light entering the water at C is said to have reached the critical angle AOC, and hence is totally reflected. By this is meant that it has gone beyond the critical point, where the law of the sines comes to an end, and reflection takes the place of refraction.

Sometimes, when walking along a road on a warm day, you may observe a curious quivering in the air just where the road seems to meet the sky, as it goes over a hill. The objects near this point appear to be distorted and to tremble, or they assume fantastic shapes. Here we have an instance of refraction caused by the heated air just above the surface of the road. The light passing through these layers of unequally-heated air is refracted unequally, and the objects that reflect the light appear distorted. In some instances the refraction may pass the critical angle, and we may see the objects apparently doubled by reflection. Warm, calm days by the sea show the same thing, when distant vessels appear repeated in the sky, or when distant land that is really below the horizon "looms" up and glimmers upon the horizon in trembling headlands. This illusion is called the mirage, and takes place when refraction exceeds the critical angle and becomes reflection.

Fill a clear glass tumbler with water, and put a spoon in it, or dip one finger in the water, and hold it above your head so that you can look into the water from below. You will find that you cannot see through the water up into the air above. The under surface of the water will appear to shine like burnished silver, and the spoon or your finger will be reflected in it, as in a beautiful mirror. This illustrates total reflection, and shows that in this case all light thrown upward through the water is reflected from its surface. Look into the tumbler from above, and it appears full of clear water. Look into it from below, and it seems as if an opaque sheet of silver rested on the water, and shut out the view of everything above.

Take a small glass tube, and roll up a piece of coloured paper or printed paper and slip it inside the tube, and then place the tube in the goblet of water. Hold the goblet in the hand near the eyes, and you can see the paper in the tube through the water. Lower the goblet till you can look down into the water from above, and the tube will appear as if made of silver, and the paper will totally disappear. To vary the experiment, lift the tube up and down in the water, and the paper will appear and disappear in the most surprising manner. This also illustrates total reflection. The light reflected from the paper passes through the glass tube into the water, and is refracted. In certain positions the light passes the critical angle, and is reflected from the outer surface of the glass tube, and fails to reach the eye. Look into the goblet from below, and there is the coloured paper pictured by total reflection on the under side of the water.

THE WATER-LENS.

Fig. 19 shows an oblong box of pine, 14 inches (35·7 centimetres) high, 6½ inches (16·5 centimetres)

FIG. 19.

square at the outside at each end, and made of thin boards, nailed or screwed together. One side is entirely open, and at the top is a round hole, 5 inches

(12·7 centimetres) in diameter. On this opening rests a hemispherical glass dish, made by cutting off the round top of a glass shade. This makes a thin glass bowl, $5\frac{1}{2}$ inches (14 centimetres) in diameter, and it rests in the hole, partly above and partly below the top of the box.

Inside the box two strips of wood are fastened, one on each side, at an angle of forty-five degrees. On these strips rests a sheet of silvered glass, $5\frac{3}{8}$ inches (13·7 centimetres) wide and $8\frac{7}{10}$ inches (21·4 centimetres) long, or just large enough to slip into the box, as shown by the dotted lines in the picture. To keep the glass from sliding out, a tack or brad may be driven in the box at the end of the mirror.

Put the heliostat in the window, and bring a full beam of sunlight into the darkened room. Then place this box on the window-seat, or on a table next to the window, with the open side toward the window, and in such a position that the beam from the heliostat will fall on the mirror. By this arrangement the light will be reflected upward through the glass bowl. Then fill the bowl with clear water, choosing the purest and cleanest that can be found. Adjust the box carefully, and see that the beam from the heliostat strikes the mirror fully, and that the reflected beam meets the bowl on every side, so that there are no shadows inside the box.

Here we have a broad beam of light passing from the air into water, and our experiments have shown us that in such an event the light may be refracted.

Hold a sheet of paper in a horizontal position just above the bowl, and you will see that it is fully lighted up by the light thrown up by the mirror through the water. Raise the paper slowly, and the circle of light on the paper will grow smaller and brighter, till it is reduced to a small dot of intense white light.

Put a match just at this bright spot of light, on the under side of the paper, and instantly it begins to burn. Touch the lighted match to the paper, and hold the burning paper beside the bowl of water, and gently blow the smoke over the water. See what a strange cone of light appears in the smoke! It is pale below, next the water, and grows brighter and brighter till the top of the cone is reached, and here it is intensely bright. Above this cone appears another, upside down, with its point touching the point of the cone beneath it. Above, on the ceiling, is a large circle of light, perhaps several feet in diameter.

Fig. 20 represents a number of rays of light entering at the left, and reflected upward from the mirror. From our experiments we learned that light passing from the air into water, and reaching the surface in a normal direction, goes straight on through the water in the same path. If it enters the water on either side of this normal, it is refracted or turned aside, and takes a new path. The greater the angle at which it enters the water, the greater the refraction. In the diagram the line in the middle represents the ray of light in the centre that meets the water at a normal, and passes straight through it and on into the air above. On either side the rays are represented as refracted, or bent out of their track, and obliged to take new paths. The greater the distance of the ray from the normal, the greater its refraction.

REFRACTION OF LIGHT.

Now, as all the rays at the same distance from the normal are refracted to the same degree, it follows that there must be a place where all these rays of refraction will meet.

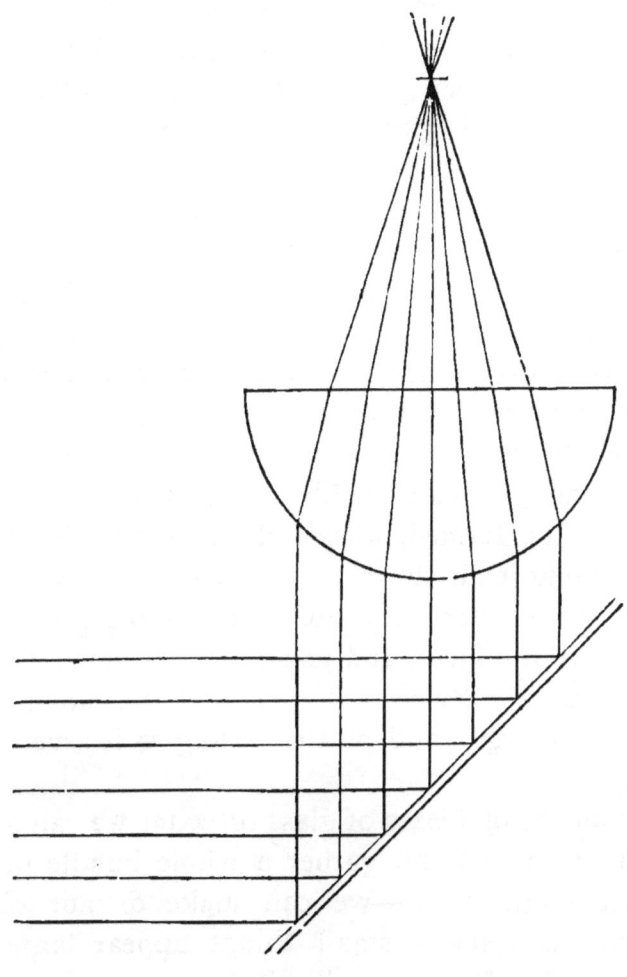

FIG. 20.

Look at the cone of light over the bowl of water, and you can see the spot where all the rays of light are concentrated. Here they meet in what is called a *focus*. You can readily remember this word, because it means a *hearth*, or *burning-place*, and we saw our match take fire just at that point. The rays of sunlight contain heat as well as light, and if we gather them altogether in a bundle, of course we shall concentrate both the heat and light. A bit of paper held in the focus glows with intensely white light, and presently begins to smoke and burn in the concentrated heat.

This bowl of water is called a lens, and, by means of refraction, we may use it to concentrate light and heat. Beyond the focus you observe the light spreads out again till it meets the ceiling, where it appears as a broad disk of light. In the diagram each ray is represented as meeting at the focus, and then all pass each other and go on in their previous directions; and you can readily see that a new cone of light will be formed, upside down, above the focus; and beyond it all the rays will spread out wider and wider the farther they go. Hold a piece of paper just above the focus, and you will see a small circle of light upon it. Raise it higher, and the circle grows larger and larger, and on the ceiling it is several feet wide.

By means of lenses of glass or water we can spread out a beam of light, gather a whole bundle of rays into a single focus—we can make distant objects appear near, make small things appear large, and large things appear small. Telescopes, microscopes, spectacles, and all kinds of optical instruments are founded on this simple law of refraction, as shown by this bowl of water.

EXPERIMENTS IN PROJECTION.

At the optician's you can purchase a small glass plano-convex lens, 3 inches (76 millimetres) in diameter, and of a focal length of about 8 inches, for perhaps less than two shillings. Such a lens is flat on one side and convex on the other, and from this it takes its name. Take this lens into a room, and close the curtains at all the windows save one. Soften a

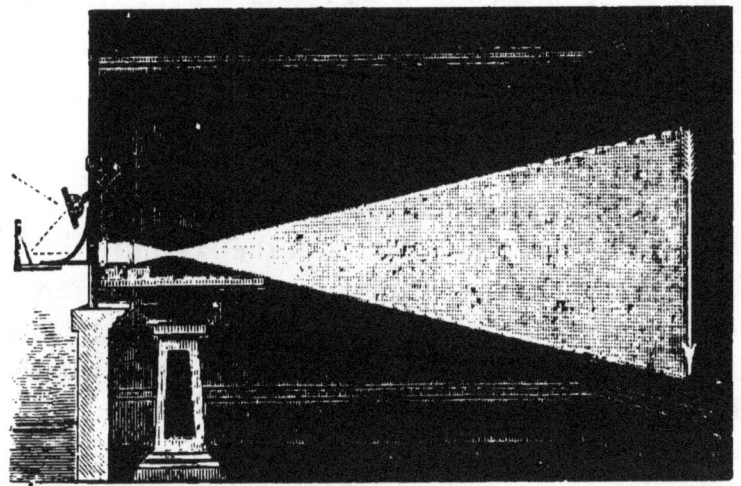

Fig. 21.

piece of wax, and stick the lens into it, so that it will serve as a handle, and then hold the lens a few inches from the wall, or, if the wall is dark-coloured, before a sheet of paper pinned upon the wall, and just opposite the window. On the wall will then appear a picture of the window, and the trees, houses, and other objects that may be seen through it. Move the lens backward or forward, and you will find a place where the image on the wall becomes distinct, and gives a miniature view of the window in its natural colours, and upside down.

Here the light from the window falls upon the lens, and is refracted to a focus. This focus consists of points, each of which is formed by the convergence of rays which come from a similar point in the window. This we call a *projection*, because the light is projected or thrown upon the wall by the lens. To understand this we must notice that every part of the window sends light into the room in every direction. Every part sends light into the entire lens, and each beam is refracted and takes a new direction beyond it, so that, ultimately, all the rays meet at the focus. In examining this projection, you will notice that the glass is quite near the wall when the focus is clear and sharp. If we measure the distance from the lens to the projection, we shall get a certain measurement. This measure we call the focal distance of the lens.

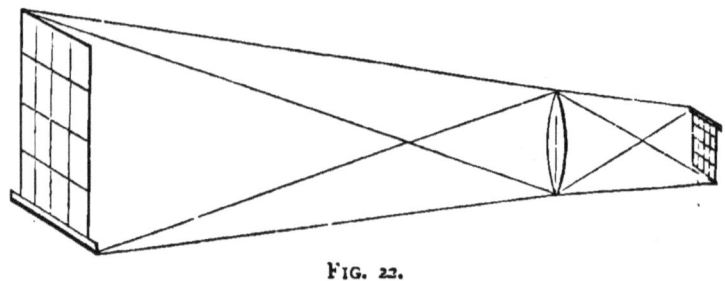

Fig. 22.

Fig. 22 represents two rays from the top of the window and two from the bottom, and shows the path they take. To draw every ray would confuse the picture, and by examining these four we can form an idea how they all travel together in a crowd and meet beyond the lens in new positions, and all closely drawn together in a focus. Those from the top of the window are refracted in one direction, those from the bottom in another, and thus they cross each other, and the projected image of the window appears to be upside down.

The picture on page 59 in this section represents the heliostat in position in a dark room. On a table in front of the instrument is the plano-convex lens, mounted on a lump of wax, fastened to a block of wood, and placed with the convex side toward the sun. The opening of the heliostat is covered by a piece of smoked glass, having a figure of an arrow drawn upon it. The light passes through the glass where the smoke was brushed away in drawing the arrow, and falls upon the lens. By refraction the beams of light form an image of the arrow upon a white screen. This screen is made of white cotton cloth, and is hung about 15 feet (4·57 metres) from the lens. The result is a large projection of the arrow, upside down, and in white on a black ground. Move the lens backward or forward slightly, and you will find a place where the projection is sharp and clear, and then the lens may be fixed there while we project other images on the screen.

This simple and inexpensive apparatus thus makes an excellent magic lantern. Common painted or photographic lantern-slides may be placed upside down at the opening of the heliostat, and will be projected on the screen clearly and distinctly, as with the best magic lanterns. Concerning the use of this lantern, and the slides that may be used in it, more may be found under the section on the water-lantern.

If it is not convenient to use a heliostat, this lantern may be used by taking a beam of sunlight, as it enters through a hole—4 inches (10 centimetres) in diameter—in the shutter, and reflecting it in a horizontal direction through the lens by means of a hand-mirror.

THE FOUNTAIN OF FIRE.

Fig. 23 represents a flat-bottomed flask, used by chemists. It has a narrow neck at the top, a flat base, and a hole at the side. This hole may be cut in the flask by means of a tube of brass one-quarter inch in diameter. This tube has a square end which is scored by two or more cross-cuts with a V-shaped file. A block of wood, having a hole of one-quarter inch in diameter, is placed against the flask; and then the tube, armed with emery and water, is inserted in this hole, and by twirling the tube in the fingers the hole in the glass is made. The tube may also be put in a lathe. In the picture the flask stands upon a shelf in front of the heliostat, and just beneath it on the floor is placed a tub or a water-pail. These few things and some pieces of coloured glass will enable us to perform a most interesting and beautiful experiment both in refraction and reflection. Place the finger over the hole in the side of the flask and fill it with water. Place the flask on the shelf so that the beam of light from the heliostat will strike the glass opposite the hole in the side.

Look at the beautiful cone of light in the water. The beam of light is refracted and brought to a focus as in our other experiments, except that here the cone is entirely under water. Study this singular

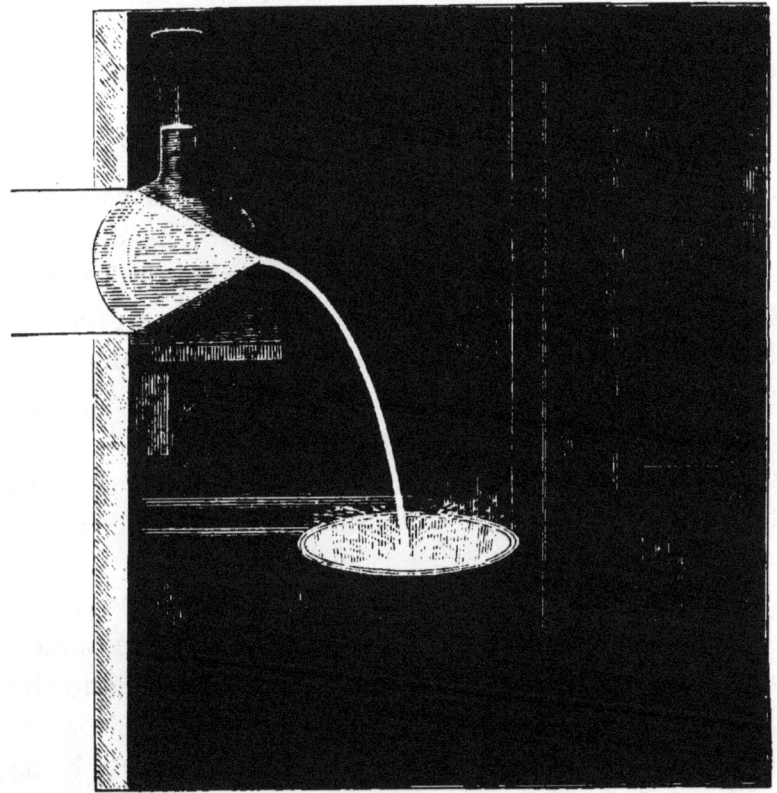

Fig. 23.

cone carefully, and adjust the flask so that the point of the cone shines on the finger at the hole in the side. When this is done, make the room as dark as possible, and then remove the finger and let the water fall in a stream into the tub on the floor.

How magical! The curving stream of water is full of light, and appears like a stream of molten iron. The spot where it falls seems touched with fire. Put your finger in the stream of water, and it is brightly illuminated. Of course, the water soon runs down, and the display stops. To prevent this, bring water in a rubber tube from the water-pipes in the house, and then regulate the supply so that the receiver may be kept full as fast as the water runs out.

Place a piece of red glass behind the flask in the beam of sunlight, and the stream of water will look like blood. Touch it, and the hand will be crimson, and the scattered drops that fall in a shower into the tub will shine like drops of red fire. Place a green or blue glass behind the flask, and the stream of water will turn green or blue, and present a most singular appearance. Hold a goblet in the stream, and it will overflow with liquid light. Flashes and sparkles of fire will appear in it, and foam over the sides, shining with brilliant light.

This beautiful experiment is as interesting as it is strange and magical, and it illustrates both refraction and total reflection. The flask makes a lens, and the falling stream of water is lighted up by the cone of light that enters it at the hole in the flask. Both the water and the light pass out of the hole together, the light inside of the water. That this is so, may be proved by permitting the water to escape, when the light will be seen shining out of the hole horizontally into the room. Why, then, does it not shine out into the room while the water is escaping? When the stream of water is flowing out, it falls in a curve into the tub on the floor. The beam of light, passing out with the water, meets its curved surface at such an angle that it is totally reflected. This beam of reflection again meets the surface of the water, and is again totally reflected. In this manner it is reflected from side to side, again and again, till it reaches the tub, and there we see it shining brightly. It is a prisoner in the water, and follows it down into the tub. When you put your hand in the falling water, you see that it is lighted brightly, and yet the stream by comparison is rather dark. If it were pure, distilled water, it would hardly be visible. As it is full of floating specks and motes, each of these reflects light, and these cause the water to appear full of light.

This fountain of fire is a charming experiment for a school, and its double lesson makes it as interesting as it is beautiful.

THE WATER-LANTERN.

Fig. 24 represents the water-lens used in the last experiment but two. The water-lens stands in the wooden box containing the mirror, and at the back of the box is a wooden slide holding a horizontal shelf at the top. This slide has a long slot cut in it, and, by means of a bolt and nut fastened at the back of the box, it can be made fast to the box in any desired position. This slide is 16 inches (40·6 centimetres) long, 5 inches (12·7 centimetres) wide, and $\frac{3}{4}$ inch (19 millimetres) thick; and the slot cut in it extends nearly the whole length. The shelf on the top is 7 inches (17·8 centimetres) long, and 5 inches (12·7 centimetres) wide, and has a hole, 3¼ inches (8·3 centimetres) in diameter, cut in the centre. The iron bolt

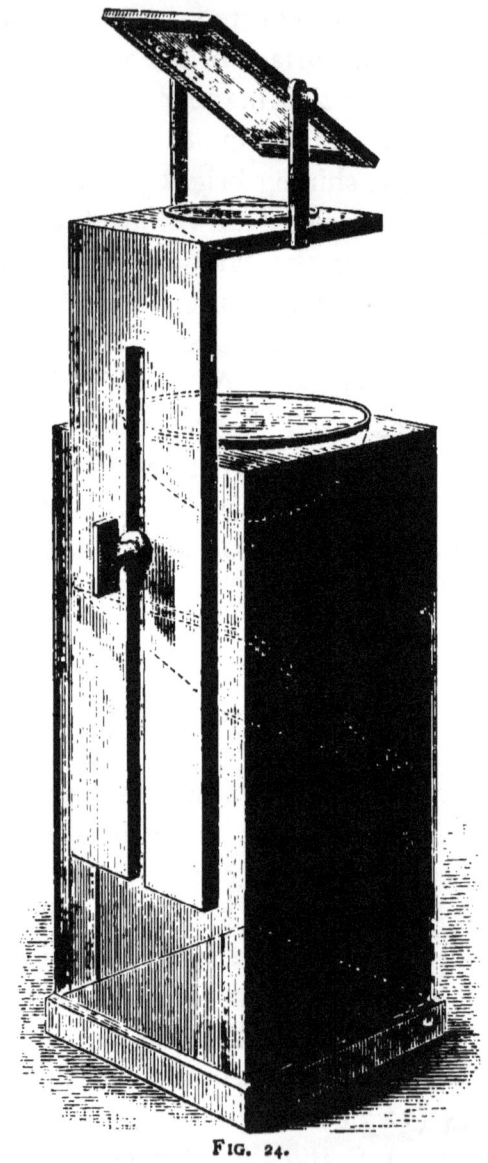

FIG. 24.

and nut must go through the back of the box, and must have a washer wide enough to cover the slot in the slide. A few inches below the bolt a block of wood is fastened to the back of the box in the slot of the slide, to serve as a guide in raising and lowering the slide carrying the lens. In the hole in the shelf rests a large watch-glass, or shallow dish, about 4 inches (10·1 centimetres) in diameter. The planoconvex lens used in our experiments in projection may be here used in place of the watch-glass. On each side of the shelf are two upright wooden arms, and between them is placed a looking-glass 7 inches (17·8 centimetres) long, and 4 inches (10·1 centimetres) wide. To hold this mirror in place, screws may be put through the top of uprights into the frame, so that it will hang suspended, and turn freely up or down.

This apparatus can be made for about 12s. 9d., the woodwork costing 4s., the two mirrors 6s., the two lenses costing 2s. 9d.; and when it is ready for work it will be a fine lantern suitable for projecting large pictures upon a screen. Place the lantern before the heliostat, so that the full beam of light will be reflected from the mirror upward through the glass bowl and the watch-glass. Fill each of these with clear water, and then place the swinging mirror at the top at an angle of 45°. Hang up a large screen of white cotton cloth or sheet in front of the lantern, and from 15 to 40 feet (4·57 to 12·20 metres) from the lantern. On this screen will appear a circle of light projected from the lantern. The sunlight from the mirror is refracted in the large water-lens and brought to a focus. It is again refracted in the small glass of water, and is reflected by the mirror on the screen. Get a piece of smoked glass, and trace upon it some letters, and then lay it on the water-lens with the top (upper side of the writing) toward the screen: immediately the letters will appear on the screen, in white on a black ground. If the projection is not distinct, loosen the nut at the back of the box, and move the wooden slide up or down till the right focus is obtained.

This water-lantern may now be used for all the work performed with ordinary magic lanterns. Place a sheet of clear glass over the large lens to keep the dust out of the water, and then lay common lantern-slides on this as in a magic lantern.

The most simple slides for such a lantern can be made by laying thin paper over engravings or drawings, and tracing the picture with lines of holes pricked with a pin. In the lantern such a paper slide will show the lines of the picture in dots of light on a dark ground. Another way is to write or draw on sheets of smoked glass. A curious effect may be made by placing the smoked glass in the lantern and writing upon it, upside down and backward, when the letters will appear to grow out in big white characters on the dark screen, and afford much amusement to all who see it. Of course, the film of smoke will easily rub off, and each scratch and finger-mark will be shown on the screen, and the work is often dirty and troublesome; but it has the advantage of being quickly done, and, if the picture is not right, it can be rubbed out and another put in its place.

A better kind of slide may be made by drawing with a needle on sheets of gelatine. Sheets of gelatine, 18 inches (45·7 centimetres) square can be bought for 1*s*. 5*d*., either pure and transparent or in a variety of colours. Lay a piece of this on an engraving, and trace the picture, drawing, map, or outline, with the point of a large needle—do not press very hard on the gelatine; a mere scratch is enough—and in the lantern every line and dot will be visible, in black upon a white or coloured ground. To preserve these sheets of gelatine, put them between sheets of glass, and bind them together with paper pasted over the edges.

Another kind of slide may be made by flowing skimmed milk over sheets of glass. When the white film of milk is dry, drawings may be traced in it with a sharp pencil or pointed stick. Another plan is to rub Castile soap over glass, and to draw on this in the same way. By this plan you can destroy the picture by rubbing on more soap, and you may then make a new picture in it.

This lantern is quite as good as the best magic lanterns. For schools, where one boy or girl wishes to show a sum in arithmetic, an example in algebra, a map, or sample of penmanship, to the whole school, the sun-lantern and piece of smoked glass, or a sheet of gelatine, will enable him to project it on a screen, so that a hundred boys and girls can see it at once.

Another interesting experiment may be made with this lantern by taking the glass cover off of the large lens, and dropping a very small chip of wood in the water. It will be pictured in gigantic size upon the screen, and curious fringes of shade will gather round it, showing where the water clings to the wood. A drop of camphor or of oil of coriander or oil of cinnamon, let fall into the water, will exhibit geometrical figures and strange motions on the screen; and a few drops of indigo or carmine ink will colour the screen blue or red, and make an excellent background for some of the pictures.

To describe all that could be done with this water-lantern and heliostat would fill a book. Having made them, you can consult other books on making projections, and find the lantern a source of amusement and instruction for hundreds of people for a very long time.

THE SOLAR MICROSCOPE.

Fig. 25 represents a common round glass flask, about 6 inches (15·3 centimetres) in diameter; a common pocket-microscope lens of 1 inch focus (costing

Fig. 25.

1*s*.), and a glass slide, carrying a microscopic object. The flask is filled with water, and is placed on a table just at the opening of the heliostat, so that the light will be refracted in it and brought to a focus. It is thus a water-lens, and may be used to bring a focus of light upon any object placed near it. Just behind this focus we place a glass slide, containing some object to be examined in a microscope. To hold this slide upright, we stick it in a mass of wax. The magnifying-glass is fastened to a bit of wax resting on a block of wood, so that it may be moved backward or forward along a strip nailed down on another block. About 15 feet (4·57 metres) from the table is placed a screen, and on this is projected a large image of the minute object on the slide. A cloth is hung over the upper part of the water-lens, to shut out the light, and all other light is excluded from the room. This apparatus makes a solar microscope, that may be used to project all kinds of microscopic objects, so that they can be exhibited before a large number of people. Tanks for holding animalcules, and all objects used in microscopes, may be placed in this solar lantern and exhibited upon a large scale, with very little trouble, and at only the expense of the flask, the pocket-lens, the screen, and the heliostat.

CHAPTER V.

THE DECOMPOSITION OF LIGHT.

Cut a vertical slit, an inch (25 millimetres) long and $\frac{1}{25}$ of an inch (1 millimetre) wide in a piece of cardboard. Make the slit with sharp, clean edges, and then fasten the cardboard over the opening in the heliostat, and a slender ribbon of light will enter the dark room. In front of this slit place a small block of wood, and on this put a lump of wax. At the optician's you can purchase for 2s. a good glass prism. Stand this upright in the wax, as in Fig. 26. Behind this, at a distance of about 15 feet (4·57 metres), hang up the screen we used in the lantern projections.

Here A is the opening in the heliostat, but somewhat exaggerated in size. The prism is at P, and S shows how the screen is placed, but gives its position much too near the prism.

On the screen will be projected a band of brilliant-coloured light, resembling the rainbow. We have seen that light may be reflected, and that it may be refracted; here we discover that, by refraction, it may be decomposed—that a single beam of white sunlight may be split into a vast number of rays, each having a colour of its own. This beautiful band of colour is called the solar spectrum. Study it carefully. It is quite impossible to count the colours, for they mingle together and merge into each other by invisible gradations, so that we cannot say where one colour begins and another ends. Yet with a little care you can make out a number of colours, that seem quite distinct. Seven colours can very easily be counted by beginning at the red, or left end of the spectrum.

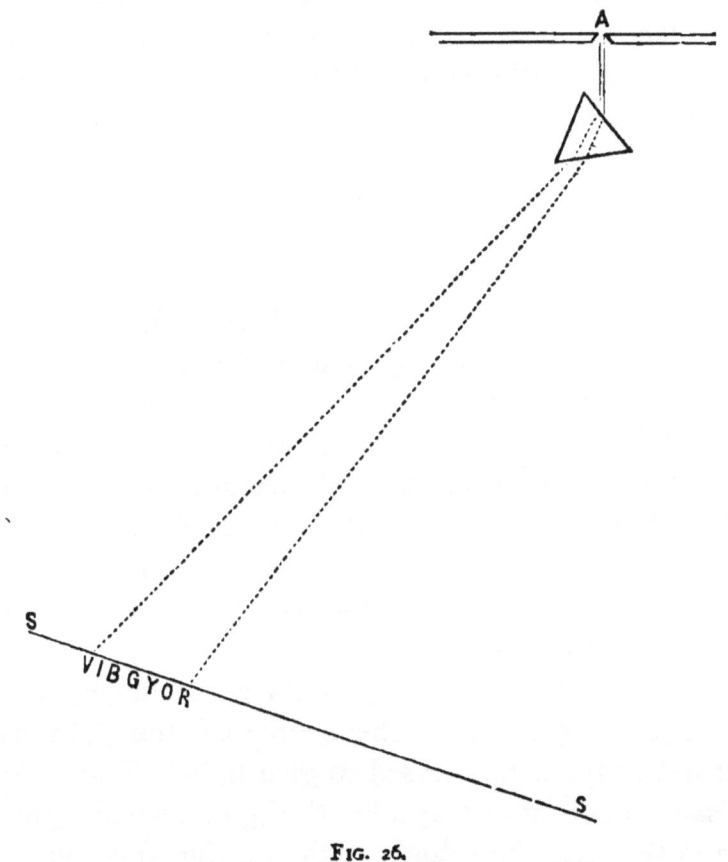

FIG. 26.

These colours are red, orange, yellow, green, blue, indigo, and violet. Some people count ten, and call them, red, orange, yellow, yellowish-green, green, bluish-green, ultramarine-blue, indigo, violet, and lavender. All these colours, and the countless shades of colours that lie between them, are in the beam of sunlight. Decompose the light by means of a prism, and they stand side by side on the screen in a beautiful band or belt. Each colour has a different degree of refraction, the refraction increasing from the red to the violet; and thus they meet the screen at different places, and we see them spread out side by side like a band or ribbon upon the screen.

To prove that this solar spectrum is the solar light decomposed, and to show that all these colours may be found in a beam of white light, place a hand-mirror in the beam of refracted light just beyond the prism. The spectrum may thus be reflected to a distant part of the room, on the wall or ceiling. Then, holding the mirror in the fingers, make it vibrate to and fro, so that the reflected spectrum will move, in the direction of its length, from side to side very quickly. At once the spectrum on the wall changes into a streak of white light, with coloured spots at each end. To understand this you must remember the common experiment of whirling a lighted stick or bit of live coal. The spot of fire changes into a ring of light. (When light falls upon the eye its effect lingers for a short time, even after the source of the light has moved away, or has ceased to give light.) The vision is said to persist or stay after the light has really gone. So in this case the colours of the moving spectrum on the wall persist or stay in the eye while they are moving to and fro, and thus one colour overlaps another, and we seem to see them all at once in one place. This mingling of every colour in the eye gives us a band or streak that appears white, and thus, indirectly, proves that all the colours of the spectrum make white, and that white light contains all these coloured lights. At the ends of this band of white are bright spots of colour. As the mirror moves backward and forward, it stops at each end of its little journey to change its direction, and here the spectrum becomes visible.

EXPERIMENTS WITH THE SOLAR SPECTRUM.

Send to the dealer in artists' materials and get a cake of red vermilion, emerald green, and aniline violet (Hoffman's violet—B. B.).[1] Get these shades, and no others, and then cut out three narrow strips of cardboard, and give one a coat of the red, one a coat of violet, and paint the other green. Take pains to give them a good thick coat, so as to hide the white paper. Study the solar spectrum on the screen carefully, and you will see that these shades of red, green, and violet are in it. When the painted strips are dry, take the red vermilion strip and hold it in the spectrum at the left or red end, and you will see that it matches the red exactly. Tip the paper backward a trifle, so that the surface of the paper will not shine or glisten in the light, and then move it slowly to the right, keeping it before the spectrum. As it passes the orange it grows dark ; in the yellow it is darker still ; opposite the green it is perfectly black. Move it to the very end, and everywhere the red strip is quite black. Place it before the red again, and its colour

* If this colour cannot be found, buy "Nuremburg Violet."

comes out clear and bright. Try the violet strip in the same way. In the same manner, the green strip is green when it is in the green part of the spectrum, and black everywhere else.

This experiment shows that green, red, and violet, are visible in green, red, and violet light, and that in light of any other colour they are invisible, and the strip of card appears to be black. Hence, an object appears of its proper colour because it absorbs all colours of the white light except its own colour, which it reflects.

Look at the spectrum closely, and you will notice that the red is at one end, the green near the middle, and violet is near the other end. Between the red and the green you will notice many shades of yellow, from deep-orange to yellowish-green, and between the green and the violet are many shades of blue, from greenish-blue to deep-indigo.

It is thought that, when we see a red light, certain nerves in the eye are affected, and convey a peculiar sensation to the brain—that we call *red*. These nerves are sensitive to red light, but are not sensitive to any other light, except in a moderate degree. Another set of nerves in the eye are peculiarly sensitive to green light, and still another set are affected by violet light. Hence the sensations caused by these three colours are called the three elementary colour sensations, and from the combinations of these sensations come all countless shades of colour. When one of these colours falls on the eye, we see it distinctly. When two—say the red and green—meet the eye, both sets of nerves are affected at once, and we get a sensation that is neither red nor green, but yellow. In the same manner, when green and violet meet in the eye, the two sets of nerves are excited, and we see not green and violet, but blue. In the same manner, if red, green, and violet light enters the eye, all the nerves are excited at once, and we see not three colours, but one, which is white.

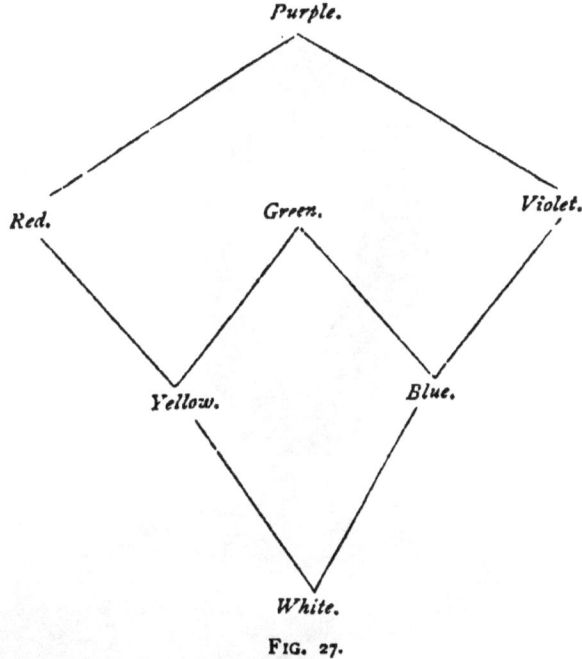

Fig. 27.

This diagram will assist us to remember the relation the colour sensations bear to each other. The red and the green combine to make yellow; the green and the violet unite to make blue; all three mingled together give us white. We may also combine red and violet light, and get purple light.

THE COLOUR-TOP.

This picture represents a common iron top, that may be found at the toy-shops. If you cannot find one exactly like it, there are others having a straight handle at the top, instead of the curved handle.

Fig. 28.

Just under the flat part of the disk are two or three round pieces of drawing-paper, and under these is a thick disk of pasteboard. Each of these has a hole in the centre, so that it can be slipped over the leg of the top. In some tops, however, it may be easier to put these disks above on the handle. Such a top as this may be made to spin in a dinner-plate on the table by winding a string round the leg, and then pulling it away with the right hand while the top is held upright by the left hand.

Get some thick drawing-paper, and cut out three disks, each 4 inches (10 centimetres) in diameter, and make a hole in the centre of each, so that it will slip over the leg of the top. Cut each disk open from the circumference to the centre with a pair of scissors. Paint one with the red vermilion, one with the emerald-green, and the other with the violet that we used in the last experiment, and then, while these are drying, make a disk of thick pasteboard, and cut a hole in the middle, so that it will slip tightly over the leg of the top.

When these are ready, take the red and green disks and hold them side by side with the cut places opposite, and slip one into the other, and then turn them round, so that the green covers the red. Then put them both on the leg of the top, and put the pasteboard disk under them, to hold them in place. Now, if you hold the top upright in a plate and make it spin, you will see a beautiful ring of green colour round the spinning top.

When it stops, take off the pasteboard, and revolve the coloured disks one on the other, so that half of the red and half of the green can be seen. Now spin them on the top, and instantly you have a ring of yellow. Move the disks again, so as to display one-quarter of the green and three-quarters of the red, and when the top spins you get a deep-orange ring.

Move them again, and let the green hide nearly all the red, and the top shows a greenish-yellow ring. In this manner you may mix the red and green in greater or less proportions, and the ring of colour on the top will exhibit new shades of yellow with every change.

In the same way, combine the green disk and the violet, and the spinning top will show a new shade of blue with every proportion in which the green and violet are mixed. Put on the red and violet disks, and let each show more or less, and shades of purple will be shown. Put on all three disks—the red, green, and violet—and arrange them so that one-third of each is shown, and the ring will be gray. Change the proportions, and you will see each time new shades of gray or white.

This is a very simple toy, but it serves to show how these three colours may be combined to produce every colour in the solar spectrum. The colour will vary very greatly, and new and beautiful shades of yellow, blue, purple, and gray, will be found at every trial. Red, green, and violet, may be tinted with other colours in the most charming manner, and hours can be filled with amusement and instruction by experimenting with this colour-top and its ever-changing coloured rings.

To exhibit the coloured rings on this top before a number of people, make a disk of stiff cardboard about 5 inches (12·7 centimetres) in diameter, and cut out three holes at equal distances from each other near the edge. Over these holes place pieces of red, green, and violet or ultramarine glass, one colour on each hole, and fasten them down with little bands of paper at the edges, and secured with mucilage. Place this disk on the colour-top, and hold it upside down just above the large lens in the water-lantern. Have the lantern prepared to give projections on the screen (see section on water-lens), and then you will see three spots of coloured light on the screen; and, by making the top spin round on the handle by means of the string, the three spots of colour will whirl round in a ring; and, if the top moves fast enough, we shall see a ring of white or grey. Cover the violet glass so as to shut out all the light, and then make the top spin, and the two spots of red and green will appear on the screen in the form of a yellow ring. In this manner all the effects exhibited by the colour-top may be projected on a large scale on the screen, and make a most interesting and beautiful exhibition that will be sure to please all who see it.

DIRECT RECOMPOSITION OF THE COLOURS OF THE SPECTRUM.

Let the spectrum fall on a mirror, and throw its reflection upon a distant part of the room. Procure a slip of looking-glass half-an-inch wide and about three inches long, and place this on the mirror in any colour of the spectrum. By tilting the slip of looking-glass, any colour can be thrown on to any other colour of the spectrum, and thus an endless variety of colours can be formed by compounding their elementary component colours.

EXPERIMENTS IN REFLECTED COLOURS.

Fig. 29 represents a flat block of wood having a short stick set up at one corner. On this stick is fastened, by means of a lump of wax, a strip of clear window-glass about 1 inch (25 millimetres) wide, and 3 inches (7·5 centimetres) long. Just behind the stick is another piece of glass of the same size fastened to the block of wood by a mass of wax. Place the instrument on a table near a window, and then sit before it with your back to the light. Cut out small bits of paper and paint one red vermilion, another emerald-green, and the third violet, as in our experiments with the colour-top. Then

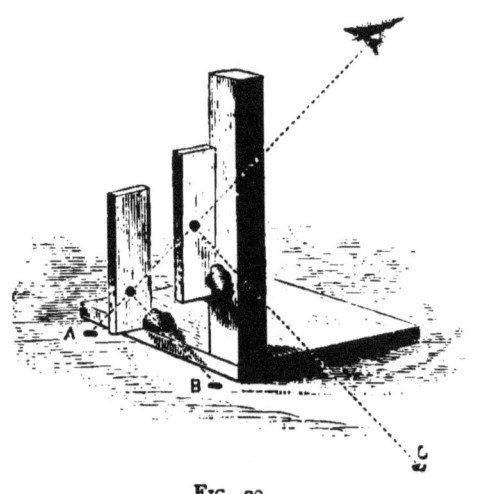

FIG. 29.

place one of these in front of the instrument, at the spot marked *C* in the drawing, and, sitting close to the instrument, look down into the glass on the stick, and you will see the bit of coloured paper reflected in the glass. Suppose this is the red piece. Then place the green piece at the ring marked *A*. On looking into the glass you can now see both the green piece reflected in the glass and the red behind it. While thus looking at both, move the one or the other till they appear to come in line, or one over the other, and then, in place of seeing a red and a green piece, you will see a single yellow piece.

Again we have a combination of colours, and we can place the red and violet or the violet and green pieces before and behind the glass, and see the colours combine precisely as in the colour-top. If it is not convenient to make this instrument, these effects can be shown with a piece of clear window-glass, by simply holding it in the hand so that one colour can be seen reflected in the glass, and the other directly through it.

With the instrument we can combine all three colours by placing them in the positions marked A, B, and C, and then looking through both glasses at once. The colour at A will be seen through both glasses, the colour at C will be seen in the upper glass and in a line with the first, and the colour at B will be seen reflected on the surface of the lower glass; and, if all three are in the right places, we shall see only one piece, and that will be white or grey.

EXPERIMENTS IN CONTRASTED COLOURS.

Cut out three pieces of drawing-paper about 2 inches (5·1 centimetres) square, and paint one red, another green, and another violet, using the paints we bought for the colour-top. If these are not at hand, cut out squares of red, green, and violet paper, and squares of yellow, pink, blue, or any other colours you can obtain. Lay a piece of black cloth on a table near the window, and then sit before it with your back to the light, and place the red square of paper on the cloth. Take a sheet of white, or still better, light-grey blotting-paper in the right hand, and hold it just above the red square in such a position that it can be quickly slipped over it, so as to hide it from sight. Now look steadily at the red square for a minute or two, and then slip the grey paper over it. In a few seconds there will appear on the grey paper a curious image just the size and shape of the red square, but of a bluish-green colour It will grow brighter quickly, and then fade away, leaving nothing but the grey paper. Put the green square on the black ground, and, after looking at it for a moment, cover it with the grey paper, and a pink image of the square will seem to shine out of the grey paper for a moment, and then fade away. Look at the violet square, and then suddenly hide it, and a pale, greenish-yellow image will be seen. Get a square of yellow paper, and repeat the experiment, and you will see a violet-blue image. In the same way try an orange square, and get a violet image; try greenish-yellow, and get a pink image.

These strange, ghostly after-images that linger after any colour has been suddenly removed, result from an action in the eye. Having looked at red till the eye becomes weary, and then having suddenly taken the red away and replaced it by a white surface, the nerves of the eye send us another sensation that we call bluish-green. The nerves sensitive to red having become fatigued, the nerves sensitive to green and violet are fresh and fully sensitive to the green and violet rays in the white light reflected from the paper.

Every colour gives a particular after image, and this image is always of a colour that is said to be complementary to it. Red is complementary to bluish-green, orange to sky-blue, yellow to violet-blue, green to pink, and so on through all the colours.

To vary this experiment, you may take a small bit of green paper, say about the size of a wafer, and lay it on the middle of a square of orange-coloured paper, and then, after looking at the two colours for a moment, hide them both with the grey paper, and the after-image will be blue, with an orange spot in the centre. Take off the green wafer and put on a blue one, and make the experiment. Put a large spot of ink in the centre of this orange square, and the blue after-image will have a grey spot in the centre. Black, we know, is the absence of colour, and on looking at the black spot the eye is not excited at all, and the blue image appears to have a hole in the middle through which we see the grey paper. If we put a white wafer on the orange square, we shall see a blue after-image with an orange spot.

These strange and curious after-images, appearing like coloured ghosts on the grey paper, may afford both amusement and instruction, and will give us the complementary colour of any colour we use in the experiments. These complementary colours, when placed side by side, always give the eye a pleasing sensation, and we say that the colours look well together.

Take a piece of red cloth or paper and hang it up before a white screen, or upon a white wall, in a dark room. Stand near the window and look at the red cloth, and you will not be able to see it. Then slowly open the window shutter, and permit a little light to enter the room. Now the red cloth looks like a black patch on a grey wall. Let in a little more light, and it turns to a deep, dark red. Gradually let in more and more light, and the deep-red cloth will change to a lighter and lighter shade, till the room is fully lighted, when it will appear in its real colour. Bring the red cloth into the full sunlight, or throw a beam from the heliostat upon it, and it assumes a still lighter shade of red. Close the shutters again, and it will change back from red to dark-red, and through every shade to perfect black. Try any other colour in the same manner, and you will produce precisely the same effects. This experiment may be tried in the night by turning the gas (or other lamp) slowly down till the light disappears, and turning it up again, while you watch the changing shades of colour as the light decreases and increases.

Here we have quite a different matter, showing that the shade or brightness of a colour depends upon the amount of light it receives. If it has plenty of light, it appears of its normal shade; if it has less light, it takes a darker shade; if more, it has a brighter shade. If, in using our colour-top, we put a bit of black cloth or paper on the colours, the ring of colour, when the top spins, will take a darker shade. If we put on a piece of white paper, we shall get rings of a lighter shade.

When the sunlight falls on any object, the object absorbs part of the light and reflects the rest. If it absorbs all the light and reflects nothing, the eye sees no reflected rays, and we say the object is black, or is invisible. If it reflects all the light, all colours enter the eye and we say it is white. If it absorbs all the rays of the spectrum except red, our eyes receive these red-reflected rays, and we call the object red. If it absorbs all the rays except red and green, the eye receives these two rays, two sets of nerves are excited at once, and we say that the object is yellow. It is in this manner that we see the things about us, and are enabled to recognise the colours in which they appear to be clothed.

CONCLUSION.

WE have now seen how light moves through air, water, and other transparent substances; we have learned something of the manner in which it may be reflected and refracted; and we have examined a few of the more simple facts about colours. Yet we have not by any means learned all that is known about light, nor have we exhausted half the capabilities of our apparatus. We have studied reflection from plane surfaces: all the wonderful effects produced by reflection from curved mirrors remain for further study. We have examined only one or two of the different kinds of lenses; and in the beautiful science of colours we have, as it were, only opened the gate into a strange and marvellous country.

You might go on for a year and still make experiments every day, and even then not reach the end. You have seen that it is not difficult to make experiments; and, should you take up other books on light and make new experiments, you would find much that would be of the greatest value and interest.

Should you learn nothing else, you will see for yourself with what skill, wisdom, and goodness, all these beneficent laws have been arranged. These things came not by chance, or of themselves. They all point to a great and wise Creator, who has given the light a pathway, and filled it with bewildering and perpetual beauty. It is the light that paints the flowers, tints the clouds, and decks the sky in blue. Everything selects its own particular colour out of the solar spectrum, and shines with all the beauty and glory of the light. No man hath counted all the glories of light, nor hath any man yet traced all its paths. It brings us strange messages from distant suns; it makes all Nature beautiful.

Having made a fair start in the art of experimenting, let us go on to new experiments in sound, heat, magnetism, electricity, and mechanics.

THE END.

www.ingramcontent.com/pod-product-compliance
Lightning Source LLC
Chambersburg PA
CBHW032245080426
42735CB00008B/1008